ANATOMY of EXERCISE FOR 50+

ANATOMY of EXERCISE FOR 50+

Hollis Lance Liebman

Firefly Books

Published by Firefly Books Ltd. 2012

First printing

Publisher Cataloging-in-Publication Data (U.S.)

Liebman, Hollis.
Anatomy of exercise for 50+ : a trainer's guide to staying fit over fifty / Hollis Liebman.
[160] p. : col. photos. ; cm.
Includes bibliographical references and index.
Summary: Instruction to learning exercises by combining photographs and lifelike anatomical drawings, to reveal the detail of which muscles are engaged in each exercise, specifically for the population age fifty and above.
ISBN-13: 978-1-77085-162-7
ISBN-13: 978-1-77085-156-6 (pbk.)
1. Exercise. 2. Muscles--Anatomy. 3. Muscle strength. I. Title.
796.44 dc23 RA781.L543 2012

Library and Archives Canada Cataloguing in Publication

A CIP record for this title is available from Library and Archives Canada

Published in the United States by
Firefly Books (U.S.) Inc.
P.O. Box 1338, Ellicott Station
Buffalo, New York 14205

Published in Canada by
Firefly Books Ltd.
66 Leek Crescent
Richmond Hill, Ontario L4B 1H1

Printed in China

This book was developed by:
Moseley Road Inc.
123 Main Street
Irvington, New York 10533

President Sean Moore
International Rights Director Karen Prince
Art Director Brian MacMullen
Editorial Director Lisa Purcell
Editor Erica Gordon-Mallin
Designers Danielle Scaramuzzo, Terasa Bernard

Photographer Jonathan Conklin Photography, Inc.
Models Elaine Altholz, Peter Vaillancourt

CONTENTS

Introduction: Working Out Smarter 8
Full-Body Anatomy 12

Flexibility Exercises 14
Swiss Ball Kneeling Lat Stretch 16
Standing Quadriceps Stretch 17
Side Lunge Stretch 18
Standing Hamstrings Stretch 20
Standing Calf Stretch 21
Triceps Stretch 22
Shoulder Stretch 23
Chest Stretch 24
Standing Biceps Stretch 25
Supine Lower-Back Stretch 26
Cat-to-Cow Stretch 27
The Bird Dog 28
Swiss Ball Abdominal Stretch 30

Endurance Exercises 32
Alternating Chest Press 34
One-Armed Row 36
Overhead Press 38
Lateral Raise 40
Standing Fly 42
Biceps Curl 44
One-Armed Triceps Kickback 46
Squat 48
Split Squat with Curl 50
Wood Chop with Resistance Band 52
One-Legged Downward Press 54

Alignment & Posture Exercises 56
Standing Abdominal Brace 58
Lying Pelvic Tilt 59
Bridge 60

Prone Limb Raises 62
Arm Raise 62
Leg Raise 63
Swiss Ball Pelvic Tilt 64
Swiss Ball Sit-to-Bridge 66
Cobra Stretch 68
Bilateral Scapular Retraction 70
Swiss Ball Arm Flexion 70
Swiss Ball Arm Extension 71

Range of Motion Exercises 72
Neck & Shoulder Exercises 74
Head Turn 74
Head Tilt 74
Back Pat & Rub 75
Arm Exercises 76
Forward Arm Reach 76
Sideways Arm Lift & Cross 77
Forearm Exercises 78
Elbow Bend & Turn 78
Wrist Bend 79
Leg Exercises 80
Knee Lift 80
In-Out Rotation 81
Ankle Rotation 81

Core-Strengthening Exercises 82
High Plank 84
Plank 86
Swiss Ball Plank with Leg Lift 88
Side Plank 90
T-Stabilization 92
Prone Cobra 94

CONTENTS continued

Swiss Ball Jackknife 96
Swiss Ball Rollout 98
Crunch 100
Reverse Crunch 102
Leg Raise 104
Seated Russian Twist 106
Bicycle Crunch 108
Swiss Ball Hip Crossover 110
Seated Arm Raise with Medicine Ball 112
Medicine Ball Over-the-Shoulder Throw 114
Big Circles with Medicine Ball 116

Toning Exercises 118
Swiss Ball Fly 120
Alternating Floor Row 122
Swiss Ball Pullover 124
Swiss Ball Seated Shoulder Press 126
Alternating Dumbbell Curl 128
Lying Triceps Extension 130
Sumo Squat 132
Lunge 134
Stiff-Legged Deadlift 136
Dumbbell Calf Raise 138
Swiss Ball Incline Chest Press 140
Dumbbell Upright Row 142
Side Lunge 144
Swiss Ball Hamstrings Curl 146

Workouts 148
Glossary 156
Credits & Acknowledgments 160

INTRODUCTION: WORKING OUT SMARTER

These days, it's a safe bet that the weight machine at the gym, the fast lane at the pool, and the hardest class at the yoga studio are populated by exercisers 50 and older. More than ever before, 50-plusers are achieving unprecedented levels of physical fitness—sometimes even getting into the best shape of their lives.

Studies show that regular exercise really does help you to live longer and more healthily, no matter when you start doing it. Whatever your age, beginning a sound fitness program and consistently following it can help you look and feel fantastic. There's a real upside to turning 50 (and beyond): Just as you know what's important as far as your career, families, and friends go, chances are you are also aware of your strengths and weaknesses. Armed with sound understanding of who you are, you also know what you want from a workout. Should you strive to improve your balance? Are you hoping to alleviate that tightness in your shoulders, while also whittling your waist? Do you love your legs, and want to keep them looking great? Drawing on mature self-knowledge, you can make better-informed choices about how, when, where, and why you exercise. And in turn, you discover smarter, more effective ways to achieve well-being.

WORKING OUT SMARTER

Consider this book your guide to working out smarter. A smart workout involves a big-picture approach, letting you hone a range of different aspects of fitness. A smart workout also addresses small problems before they become bigger ones. And even if none of your muscles, tendons, ligaments, and joints are feeling strained, sore, or inflexible, exercise can do much to keep them that way.

The following pages will give you a comprehensive exercise program, devised with attention to your whole-body anatomy. The first group of exercises focus on flexibility, while the next ones improve your endurance. You'll also find exercises focused on posture and alignment, others that target your range of motion, some that strengthen your core to prevent injury, and others that tone. Performed together, these exercises will improve not only the aesthetics of your body (how it looks), but also its functionality (how it performs).

Alongside each major exercise, you will see an illustration showing the muscles challenged. As you work out, visualize the muscles that are being engaged as your body becomes stronger, tighter, and leaner.

YOUR FITNESS TOOLS

Some of these exercises incorporate equipment—all reasonably small tools that add variety and challenge to the at-home workout.

Several of the toning exercises call for small hand weights or adjustable dumbbells. You can start with very light, 5-pound weights (or even lighter substitutes, such as unopened food cans or water bottles), and then work your way up to heavier ones.

You will also see two kinds of ball: a small, weighted medicine ball, which is used like a free weight, and a larger Swiss ball. Also known as an exercise ball, fitness ball, body ball, or balance ball, this heavy-duty inflatable ball is available in a variety of sizes, with diameters ranging from 18 to 30 inches. Be sure to find the best size for your height and weight. Swiss balls are excellent fitness aids that really work your core. Because they are unstable, you must constantly adjust your balance while performing a movement, which helps you improve your overall sense of balance and your flexibility.

You will see two types of resistance

WORKING WITH BANDS

As simple as they are, resistance bands are amazing pieces of fitness equipment, which effectively tone and strengthen your entire body. You may wonder, though, just how they work.

Bands act in a similar way to free weights, but unlike weights, which rely on gravity to determine the resistance, bands use constant tension—supplied by your muscles—to add resistance to your movements and improve your overall coordination.

Resistance bands are made from elastic rubber and usually come in two forms: wide and flat or tubular with handles. You can use either type to achieve the same results, and both flat and tubular bands come in several levels of tension, from very stretchy to very taut.

Whether you are a novice or an expert, you can add resistance bands to your fitness collection. They are relatively inexpensive and wonderfully lightweight—and they are ideal for the traveler, taking up next to no room in even the smallest carryon suitcase.

Tubular resistance band

Swiss ball

bands, one with handles and one without, both useful in adding resistance to your endurance regime.

With each exercise, you will also see a suggested number of repetitions; this number represents a goal to aim for over time, so don't worry if you can't complete so many reps right away.

WHEN AND WHERE?

At any age, working out can be a struggle—and not necessarily due to the running, lifting, and stretching involved. Rather, we struggle because we have other things going on; our lives are so busy and demanding that it's hard to find time to devote to exercising on a regular basis.

That's where the home workout comes in. Once you've made time and space for your workout, find a timetable that works for you. For example, set aside just 10 minutes a day, two or three times a week. Often, earmarking a regular time (say, after dinner) and space (the living room, for instance) encourages a consistent workout schedule. Just as you can build up the weight on your dumbbells, so too can you build up the hours per week you spend exercising. As you get more comfortable with the workout, devote more time to it to see faster, better results.

You may wish to start by flipping to the section that addresses your goal, whether you want to limber up, tone your arms, or stabilize your core muscles. Dip into this book over time, and don't be afraid to try something new; you may find an exercise that challenges you in a new way or discover that you're weak or strong in an area you never knew existed. Pay attention to what you feel your limits are, and then work toward exceeding them.

A NEW START

We all want to win on the baseball field, keep up with the kids (and maybe even grandkids), and look better than ever—yes, really—in that beautiful pair of designer jeans. At 50+, you are unprecedentedly well-equipped to make it happen. Here's to improving your whole-body anatomy through exercise, and to getting better each and every day.

FULL-BODY ANATOMY

FRONT

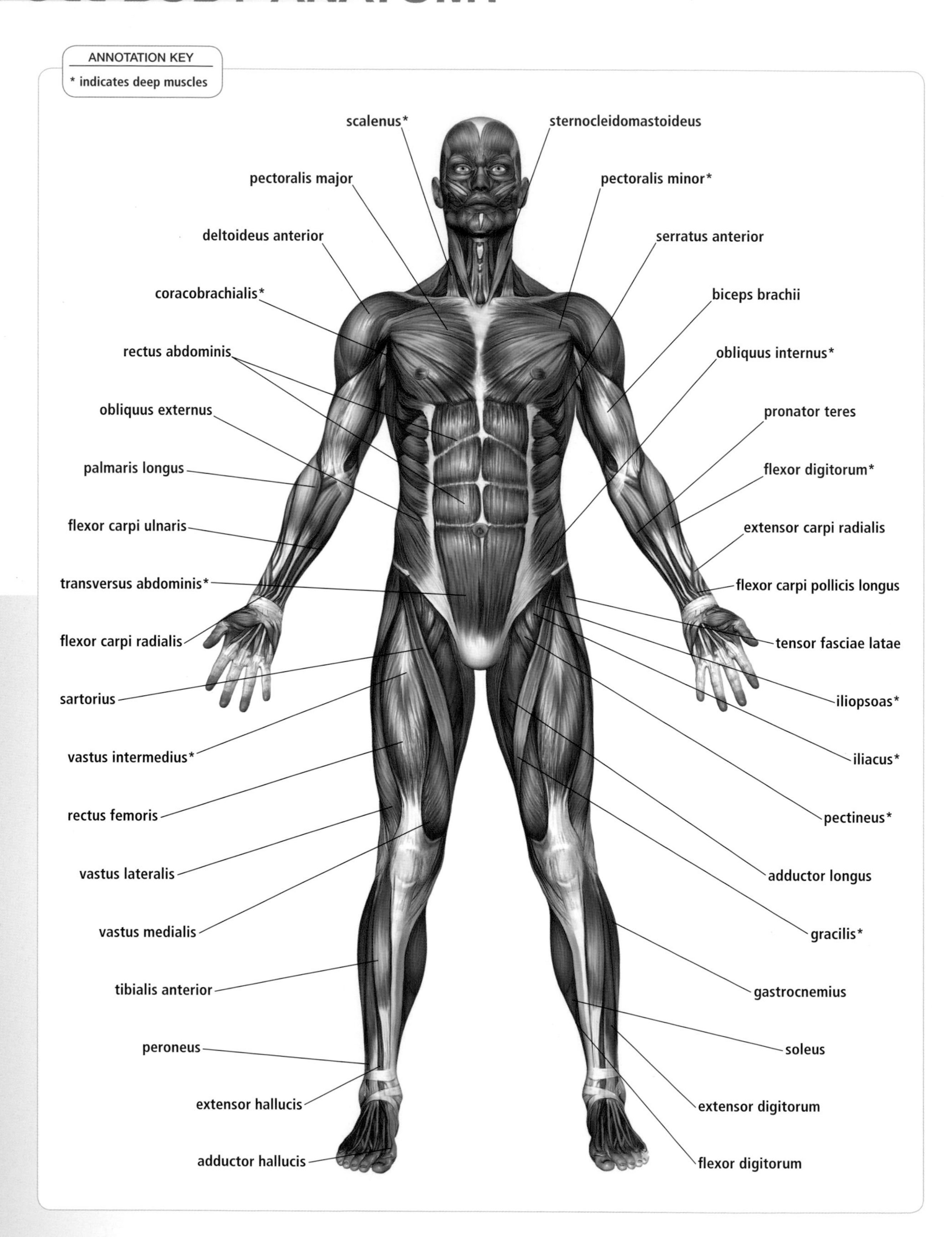

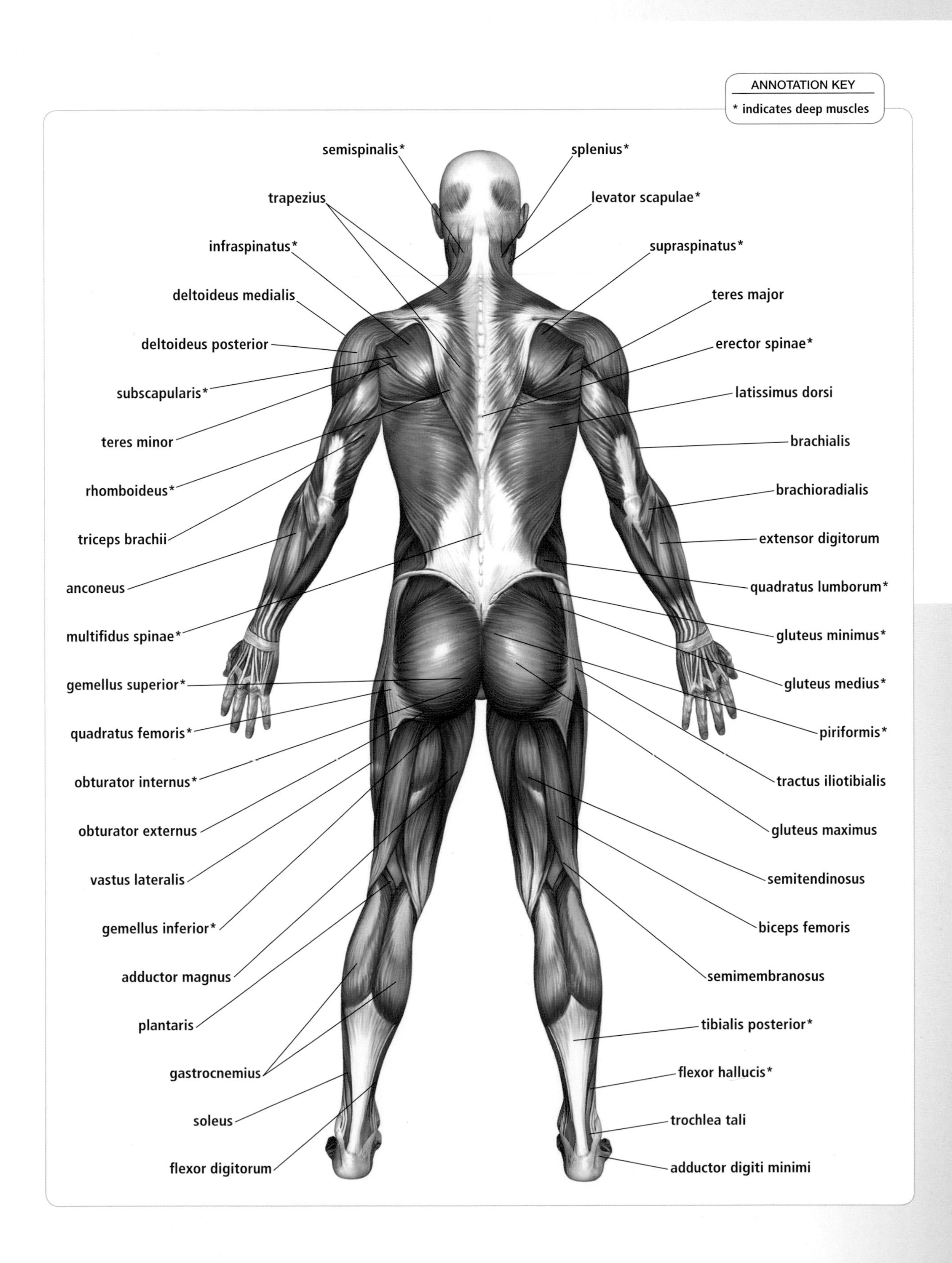
ANNOTATION KEY
* indicates deep muscles
semispinalis*
splenius*
trapezius
levator scapulae*
infraspinatus*
supraspinatus*
deltoideus medialis
teres major
deltoideus posterior
erector spinae*
subscapularis*
latissimus dorsi
teres minor
brachialis
rhomboideus*
brachioradialis
triceps brachii
extensor digitorum
anconeus
quadratus lumborum*
multifidus spinae*
gluteus minimus*
gemellus superior*
gluteus medius*
quadratus femoris*
piriformis*
obturator internus*
tractus iliotibialis
obturator externus
gluteus maximus
vastus lateralis
semitendinosus
gemellus inferior*
biceps femoris
adductor magnus
semimembranosus
plantaris
tibialis posterior*
gastrocnemius
flexor hallucis*
soleus
trochlea tali
flexor digitorum
adductor digiti minimi

FLEXIBILITY EXERCISES

Flexibility improves our performance when we play sports, exercise at home, and bend, lift, and reach our way through daily life. Most of us are born with natural flexibility in certain areas but face challenges in others. There's good news: flexibility can always be honed, over time and with practice. As you perform the following exercises, pay attention to your body's current comfort zone—and then try to transcend it. Little by little, you'll feel big results.

SWISS BALL KNEELING LAT STRETCH

1. Kneel on all fours in front of your Swiss ball. Extend one arm, placing your hand on the ball. Rest your other hand on the floor.

2. Lean back onto your heels until you feel a deep stretch in large muscles on either side of your back. Hold for 30 seconds.

3. Switch arms, and repeat. Complete three 30-second holds per arm.

TARGETS
- Back

LEVEL
- Beginner

BENEFITS
- Helps to keep back muscles flexible

AVOID IF YOU HAVE . . .
- Lower-back issues

BEST FOR

- **latissimus dorsi**
- **erector spinae**

DON'T
- Allow your torso to twist.
- Arch your neck.

DO
- Keep your arm fully extended on the ball.
- Face the floor throughout the stretch.

ANNOTATION KEY
Bold text indicates target muscles
Gray text indicates other working muscles
* indicates deep muscles

infraspinatus*
supraspinatus*
deltoideus posterior
subscapularis*
triceps brachii
teres minor
latissimus dorsi
erector spinae*

STANDING QUADRICEPS STRETCH

1. Stand with one leg bent, grasping the ankle with your same-side hand.
2. Feel the stretch in the large muscles at the front of your thigh as you bring your heel toward your buttocks. Hold for 30 seconds.
3. Switch legs and repeat. Perform three 30-second holds on each leg.

DON'T
- Lean forward.
- Arch your back.
- Hunch your shoulders.

BEST FOR

- rectus femoris
- vastus lateralis
- vastus intermedius
- vastus medialis

ANNOTATION KEY
Bold text indicates target muscles
Gray text indicates other working muscles
* indicates deep muscles

DO
- Keep your torso upright.
- Pull your foot toward your body gently, stretching only as far as you feel comfortable.
- Gaze forward.

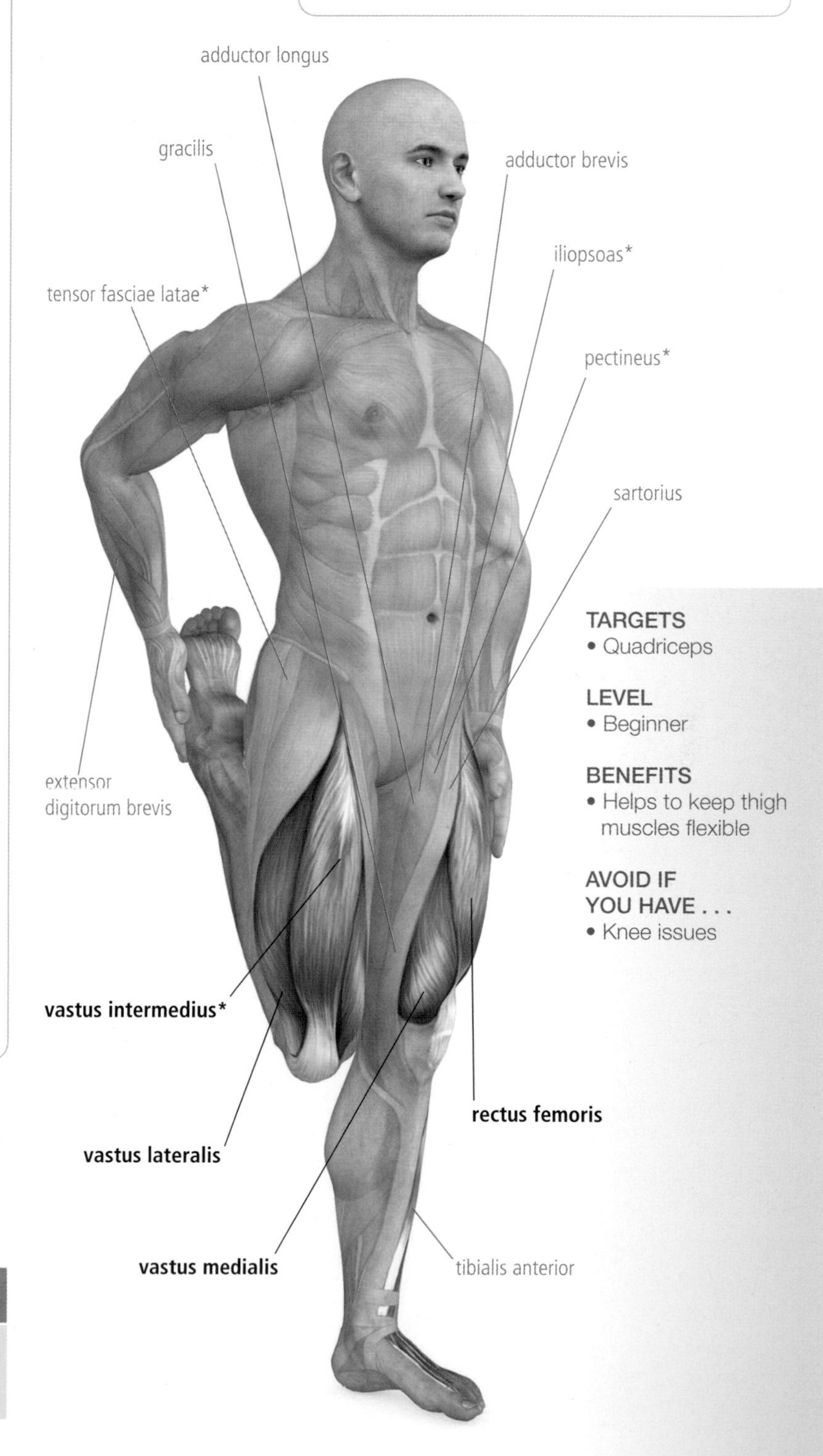

TARGETS
- Quadriceps

LEVEL
- Beginner

BENEFITS
- Helps to keep thigh muscles flexible

AVOID IF YOU HAVE . . .
- Knee issues

SIDE LUNGE STRETCH

1. Begin bent over, with your feet planted far apart. Allow your arms to dangle toward the floor in front of you.

2. Bend one knee while keeping the other leg straight. Place most of your weight on your bent leg, feeling the stretch in your lengthened leg. Hold for 30 seconds.

3. Repeat on the other side. Complete three 30-second holds on each leg.

BEST FOR

- **rectus femoris**
- **vastus lateralis**
- **vastus intermedius**
- **vastus medialis**

DON'T
- Allow your torso to twist.
- Curve your back forward.
- Arch your back or your neck.

TARGETS
- Buttocks
- Inner thighs
- Quadriceps

LEVEL
- Beginner

BENEFITS
- Helps to keep leg muscles flexible

AVOID IF YOU HAVE . . .
- Knee issues

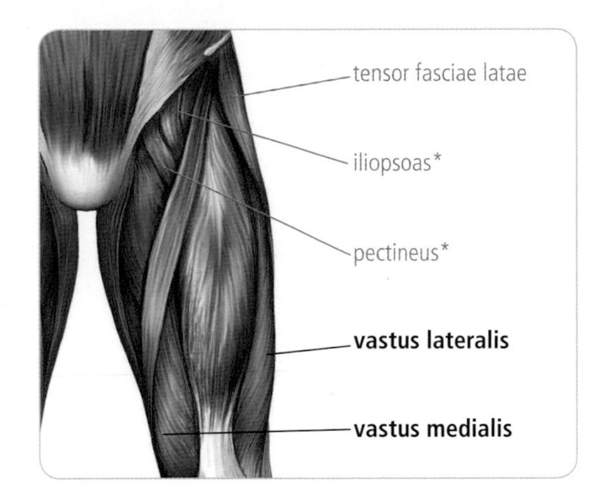

ANNOTATION KEY

Bold text indicates target muscles

Gray text indicates other working muscles

* indicates deep muscles

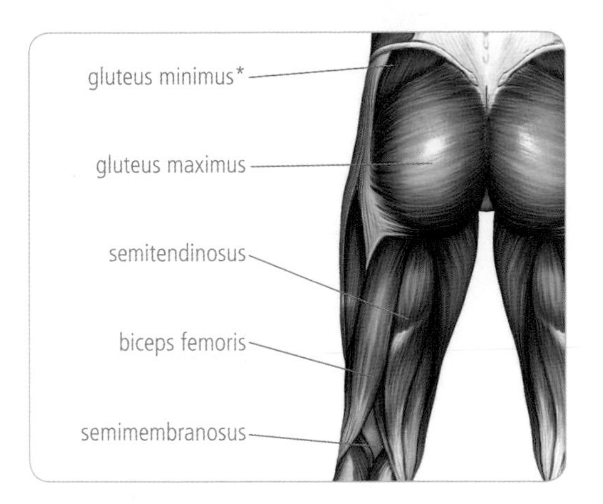

DO

- Extend your leg fully while in the stretch position.
- Gaze toward the floor throughout the stretch.
- Keep your feet flat on the floor.

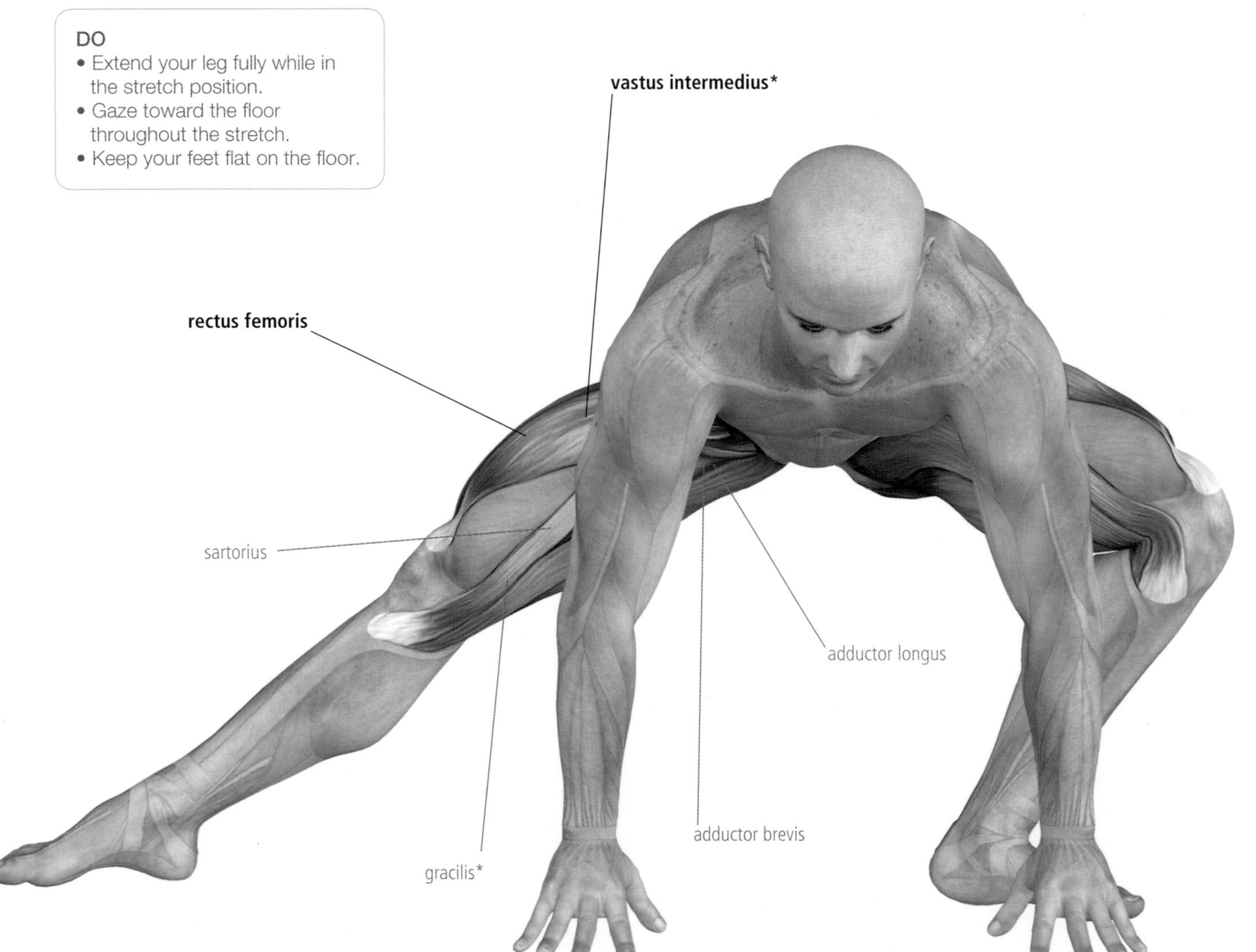

STANDING HAMSTRINGS STRETCH

BEST FOR

- **biceps femoris**
- **semitendinosus**
- **semimembranosus**

TARGETS
- Hamstrings

LEVEL
- Beginner

BENEFITS
- Helps to keep hamstring muscles flexible

AVOID IF YOU HAVE . . .
- Lower-back issues
- Knee issues

1. Stand with one leg bent and the other extended in front of you with the heel on the floor.

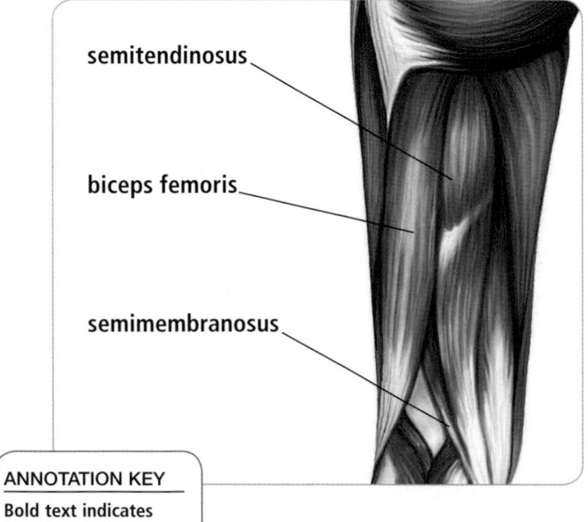

ANNOTATION KEY
Bold text indicates target muscles

DO
- Keep your front leg straight.
- Flex the foot of your front leg as you stretch.

2. Lean over your extended leg, resting both hands above your knee. Place the majority of your body weight on your front heel while feeling the stretch in the back of your thigh. Hold for 30 seconds.

3. Switch sides and repeat. Complete three 30-second holds on each leg.

DON'T
- Allow your back to arch or round forward.
- Hunch your shoulders.

STANDING CALF STRETCH

1. Stand with one foot in front of the other, with the front leg bent. With a straight back, lean over your front leg, resting both hands above the knee.
2. Place the majority of your body weight on your front heel as you feel the stretch in the calf muscle of your back leg. Hold for 30 seconds.
3. Switch sides and repeat. Complete three 30-second holds per leg.

TARGETS
- Calves

LEVEL
- Intermediate

BENEFITS
- Helps to keep calf muscles flexible

AVOID IF YOU HAVE . . .
- Knee issues

DON'T
- Raise your back heel off the floor.
- Arch or round your back.

BEST FOR
- **gastrocnemius**

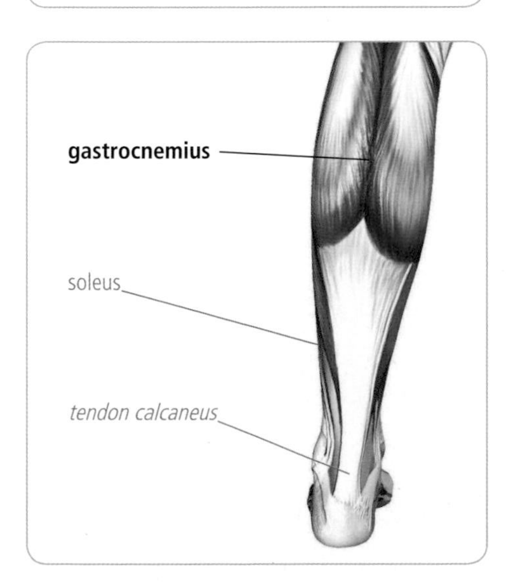

DO
- Keep both feet flat on the floor.
- Stretch with good alignment so that your back leg and your spine form a straight line

ANNOTATION KEY
Bold text indicates target muscles
Gray text indicates other working muscles
Italic text indicates tendons

TRICEPS STRETCH

FLEXIBILITY

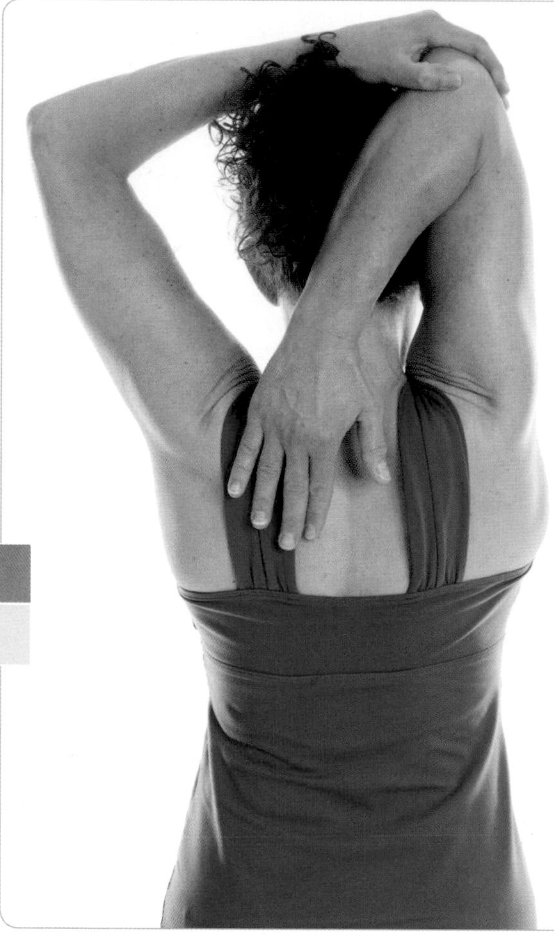

1. Stand with arms extended over your head. Bend both arms so that one hand grasps the opposite elbow.
2. Gently pull the elbow of your stretching arm as you feel the stretch in your shoulder. Hold for 30 seconds.
3. Release, then switch sides and repeat. Perform three 30-second holds per shoulder.

BEST FOR

- triceps brachii

DO

- Keep your stretching arm bent at the elbow.

DON'T

- Release your grip on the elbow of your stretching arm.
- Pull too hard on your stretching arm.
- Hunch your shoulders.

TARGETS

- Fronts of shoulders

LEVEL

- Intermediate

BENEFITS

- Helps to keep shoulders flexible

AVOID IF YOU HAVE . . .

- Shoulder issues

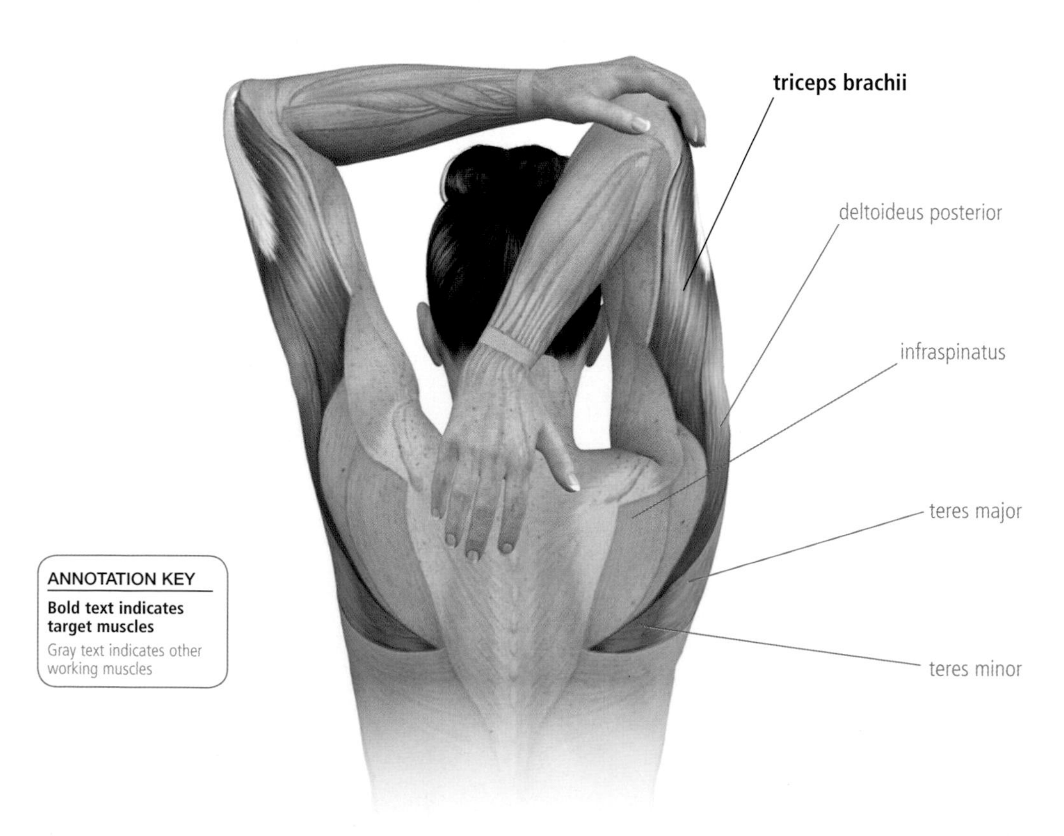

ANNOTATION KEY

Bold text indicates target muscles

Gray text indicates other working muscles

SHOULDER STRETCH

DO
- Keep your shoulders down.
- Gaze forward.
- Maintain a soft bend in your knees.

DON'T
- Allow your torso to twist.

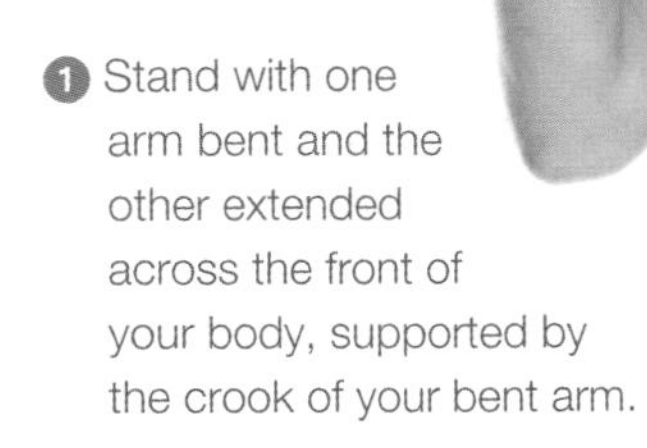

1. Stand with one arm bent and the other extended across the front of your body, supported by the crook of your bent arm.
2. Gently deepen the bend in your arm as you stretch your other, extended arm. Hold for 30 seconds, feeling the stretch in your shoulder.
3. Switch sides and repeat. Complete three 30-second holds on each arm.

ANNOTATION KEY
Bold text indicates target muscles
Gray text indicates other working muscles

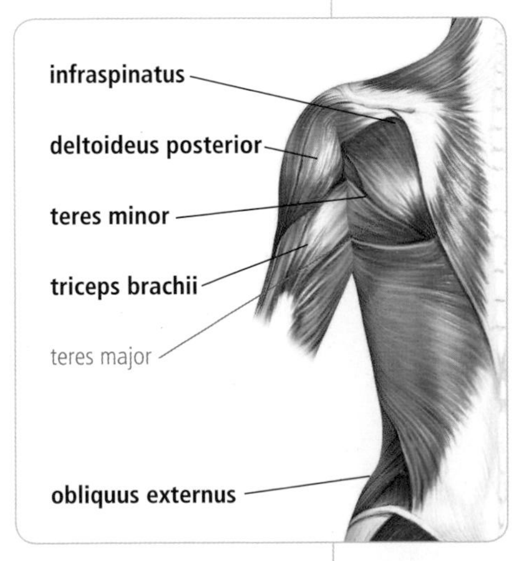

BEST FOR
- **triceps brachii**
- **deltoideus posterior**
- **infraspinatus**
- **teres minor**
- **obliquus externus**

TARGETS
- Backs of shoulders

LEVEL
- Intermediate

BENEFITS
- Helps to keep shoulders flexible

AVOID IF YOU HAVE . . .
- Shoulder issues

CHEST STRETCH

1. Stand with your hands behind your head, with fingers interlocked. Your elbows should be pointing outward.
2. Draw your elbows back as you feel the stretch in your chest. Hold for 30 seconds.
3. Bring your elbows back to starting position, and repeat. Perform three 30-second holds.

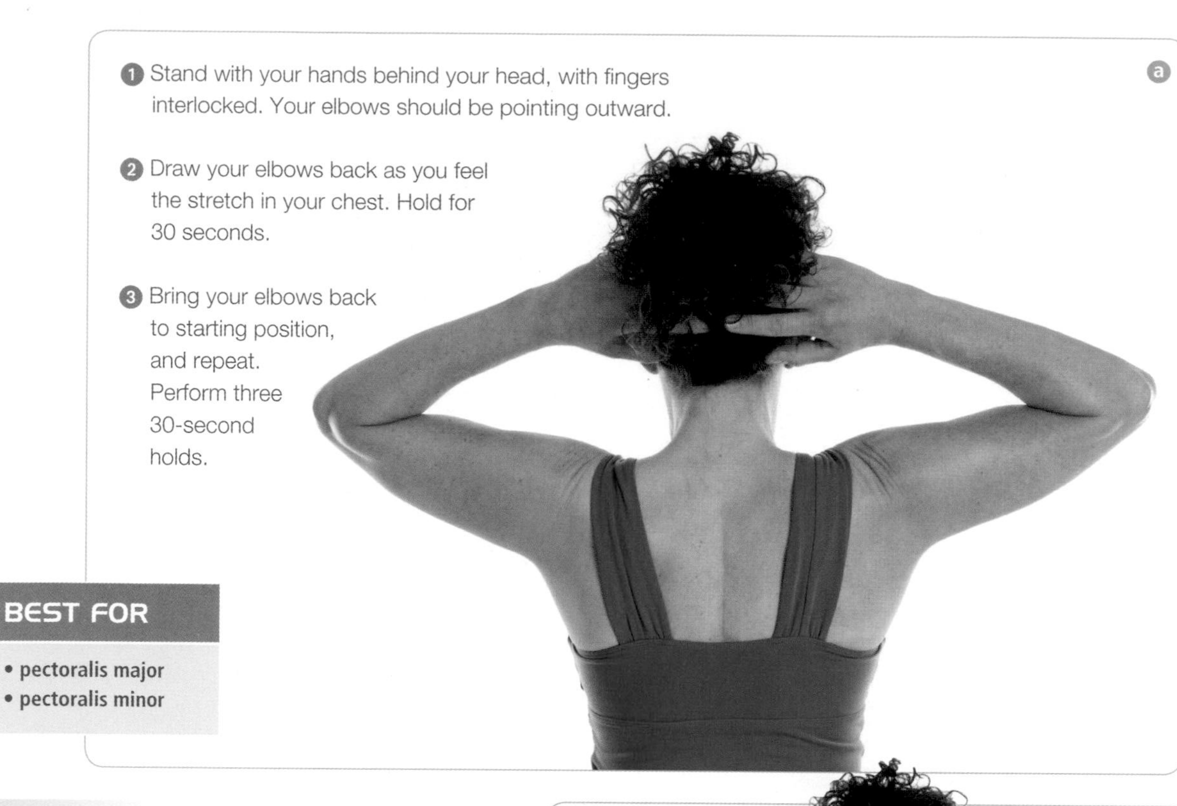

BEST FOR

- **pectoralis major**
- **pectoralis minor**

TARGETS
- Chest

LEVEL
- Beginner

BENEFITS
- Helps to keep chest muscles flexible

AVOID IF YOU HAVE . . .
- Shoulder issues

ANNOTATION KEY

Bold text indicates target muscles

Gray text indicates other working muscles

* indicates deep muscles

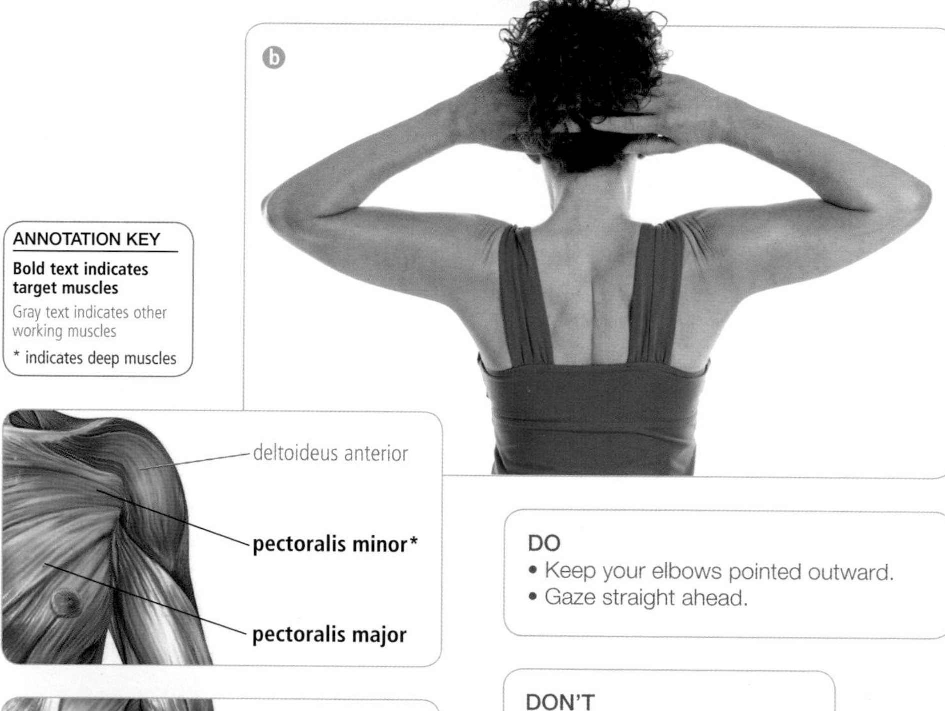

DO
- Keep your elbows pointed outward.
- Gaze straight ahead.

DON'T
- Turn your elbows inward.
- Hunch your shoulders.
- Arch your back or neck.

STANDING BICEPS STRETCH

1. Stand with your hands behind your back, clasping them together with fingers interlaced.

2. Lift your arms a few inches away from your body, allowing your fingers to stretch. Hold for 30 seconds.

3. Relax and repeat, performing three 30-second holds.

BEST FOR

- **biceps brachii**

ANNOTATION KEY

Bold text indicates target muscles

Gray text indicates other working muscles

* indicates deep muscles

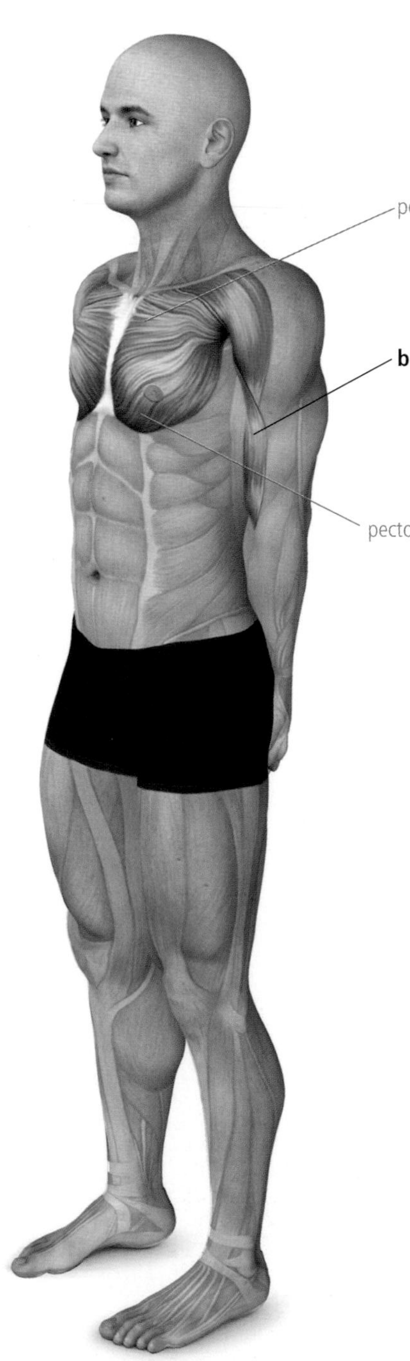

TARGETS
- Biceps

LEVEL
- Intermediate

BENEFITS
- Helps to keep biceps flexible

AVOID IF YOU HAVE . . .
- Shoulder issues

DO
- Keep your shoulders down.
- Keep your torso still.

DON'T
- Allow your hands to unclasp.
- Arch your back.
- Hunch your shoulders.
- Lift your arms uncomfortably high.

SUPINE LOWER-BACK STRETCH

FLEXIBILITY

a

1. Lie on your back with your arms and legs extended, arms angled slightly away from your body.

b

2. Bend your legs, hugging them to your body with your hands clasped around your knees. Slowly pull your knees toward your chest, feeling the stretch in your lower back. Hold for 30 seconds.

3. Relax and repeat for an additional 30 seconds.

TARGETS
- Buttocks
- Lower back

LEVEL
- Beginner

BENEFITS
- Helps to keep lower-back and gluteal muscles flexible

AVOID IF YOU HAVE . . .
- Severe back pain
- Numbness or tingling in the lower extremities

DO
- Keep your knees and feet together.

DON'T
- Raise your head off the floor.

BEST FOR
- **erector spinae**

ANNOTATION KEY
Bold text indicates target muscles
Gray text indicates other working muscles
* indicates deep muscles

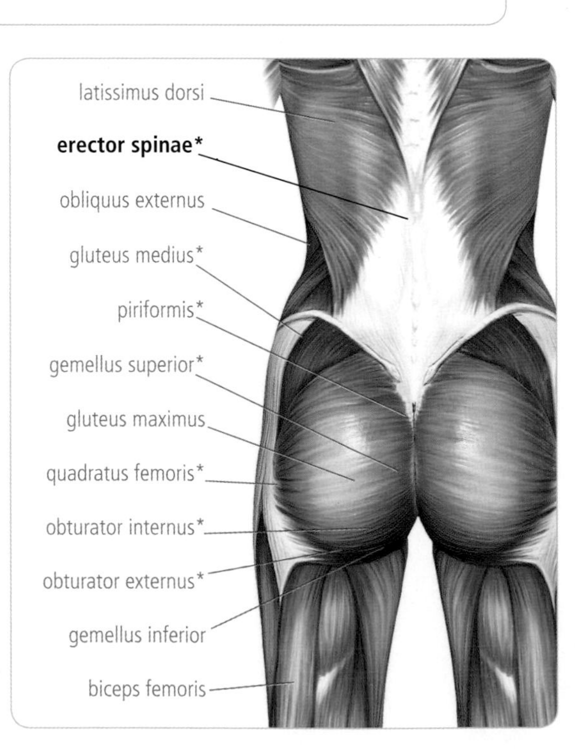

CAT-TO-COW STRETCH

1. Kneel on all fours, with your back straight.
2. Slowly curl your spine, tucking your chin slightly as you take on the position of a stretching cat. Hold for 15 seconds.
3. Transition to Cow Stretch by releasing the curve of your spine and then moving into a slight arch. Hold for 15 seconds.
4. Relax and repeat, completing three 30-second repetitions.

DON'T
- Curl or arch your back too abruptly.

ANNOTATION KEY
Bold text indicates target muscles
Gray text indicates other working muscles
* indicates deep muscles

DO
- Stretch slowly and with control.
- Keep your hands and feet planted throughout the stretch.
- Lift your chin while your spine is arched.
- Start the movement of your spine in your tailbone.

BEST FOR
- **erector spinae**

TARGETS
- Back

LEVEL
- Beginner

BENEFITS
- Helps to keep back flexible

AVOID IF YOU HAVE . . .
- Knee injury

THE BIRD DOG

1. Begin on your hands and knees, with your back straight and your abs pulled in.

DO
- Move slowly and with control.
- Keep your neck relaxed and your gaze toward the floor.
- Tuck your chin slightly while contracting your arm and leg inward.
- Keep your abs pulled in throughout the stretch.

2. Keeping your torso stable and your abs engaged, contract one of your arms and the opposite leg into your body.

3. Extend that arm and leg outward. Hold for up to 15 seconds.

DON'T
- Arch your back while your arm and leg are extended.
- Twist your torso.
- Arch your neck.

TARGETS
- Back
- Buttocks
- Stomach

LEVEL
- Intermediate

BENEFITS
- Stretches and tones arms, legs, and abdominals

AVOID IF YOU HAVE . . .
- Wrist pain
- Lower-back pain
- Knee injury

4. Return to the starting position, switch sides, and repeat.

BEST FOR
- **erector spinae**
- **rectus abdominis**
- **transversus abdominis**
- **gluteus maximus**

MODIFICATION

Harder: From starting position, extend one arm and the opposite leg straight out to the side. Hold, release, and repeat on the other side.

Harder: Begin facedown on top of your Swiss ball, with your core and upper thighs supported. Your hands should be on the floor and your legs extended with toes on the floor. Slowly and with control, lift one arm and the opposite leg upward, keeping the rest of your body stable. Hold, release, and repeat on the other side.

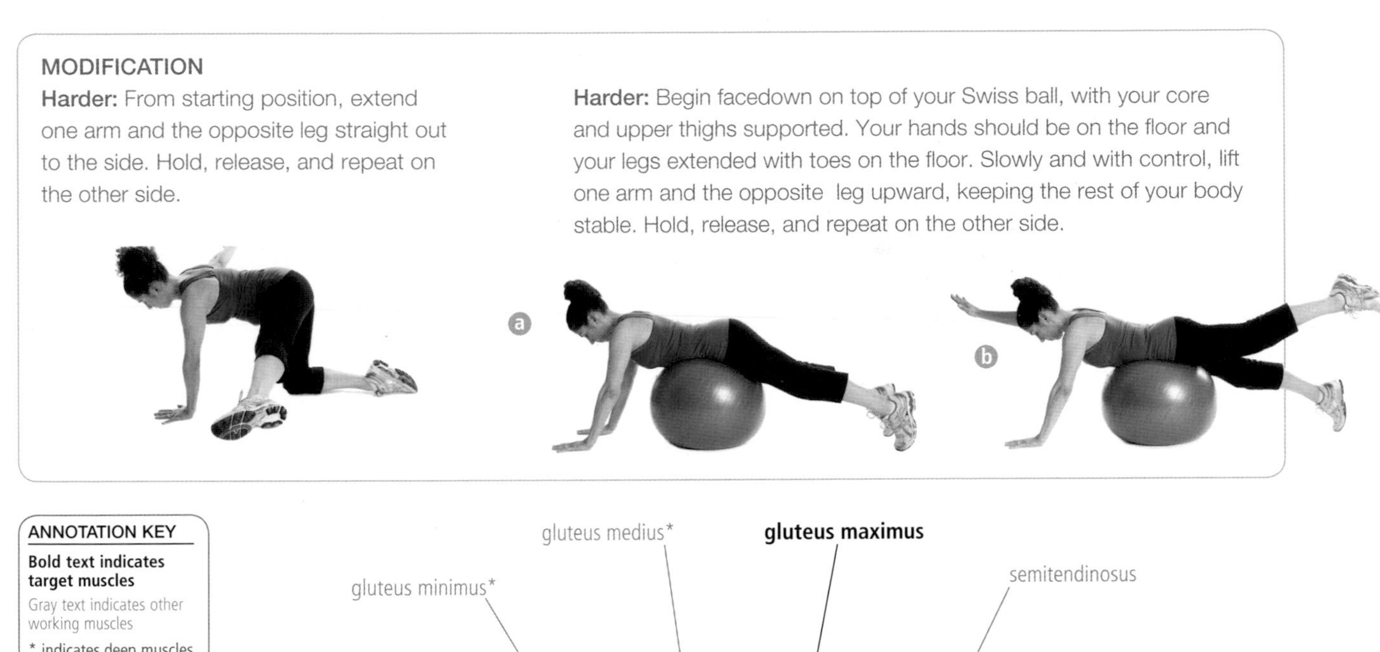

ANNOTATION KEY

Bold text indicates target muscles

Gray text indicates other working muscles

* indicates deep muscles

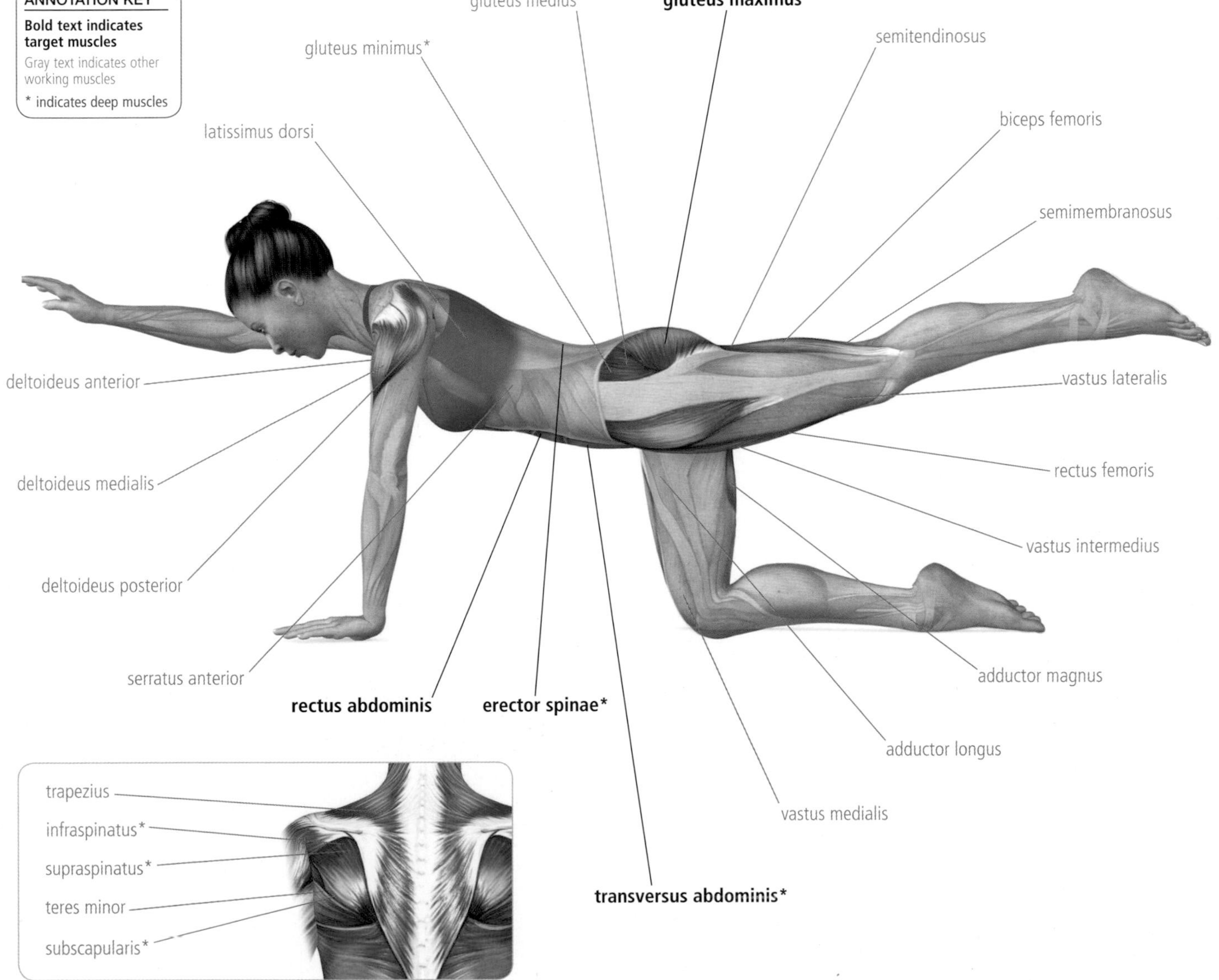

SWISS BALL ABDOMINAL STRETCH

FLEXIBILITY

1. Lie with the Swiss ball beneath your back. Your arms should be extended diagonally behind your head, your feet planted shoulder-width apart, and your knees bent.

TARGETS
- Upper abdominals

LEVEL
- Beginner

BENEFITS
- Stretches upper abdominals

AVOID IF YOU HAVE . . .
- Lower-back pain
- Balancing difficulties

DO
- Keep your torso planted on the ball.

BEST FOR
- **rectus abdominis**

2. Slowly reach downward with your arms until your fingers rest on the floor.

DON'T
- Overextend your pelvis as you raise it.

C

3 While keeping your lower back on the ball, lower your hips and stretch your abdominals toward the ceiling. Hold for 30 seconds.

4 Relax and repeat for an additional 30 seconds.

ANNOTATION KEY

Bold text indicates target muscles

Gray text indicates other working muscles

* indicates deep muscles

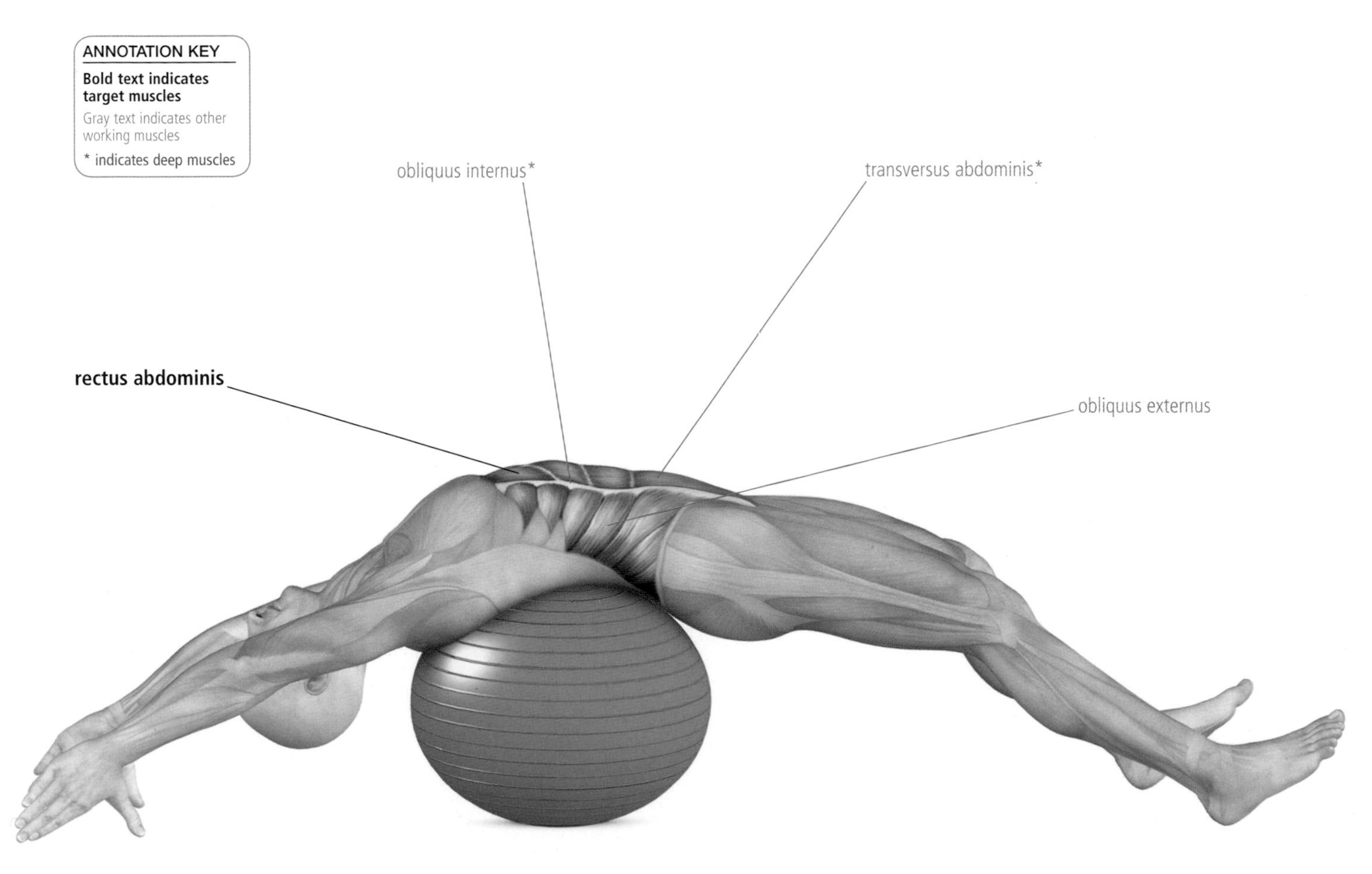

ENDURANCE EXERCISES

Endurance goes beyond isolated bursts of strength; in day-to-day life, sustained, powerful use of the cardiovascular system makes us feel stronger and perform better. The following exercises will help you to maintain and improve endurance, training your muscles to work harmoniously for longer and longer. Using resistance bands will challenge your muscles and bring about extra toning benefits.

ALTERNATING CHEST PRESS

1. Run a resistance band around a sturdy, stable object such as a pole or column. Stand facing away from the object, holding both ends of the resistance band in front of your chest.

2. Extend one arm straight in front of you to full lockout position, keeping the other arm steady.

3. With control, bring the arm back to starting position. Repeat with the other arm, completing three sets of 15 repetitions per arm.

TARGETS
- Chest
- Core
- Shoulders
- Triceps

LEVEL
- Beginner

BENEFITS
- Strengthens and tones pectorals
- Stabilizes core

AVOID IF YOU HAVE . . .
- Shoulder issues

BEST FOR

- **pectoralis major**

DO
- Keep one arm motionless as you extend the other to lockout.
- Maintain a stable torso.
- Keep your feet in place as you extend your arm.
- Engage your abdominal muscles throughout the exercise.
- Keep your arms level with your shoulders.

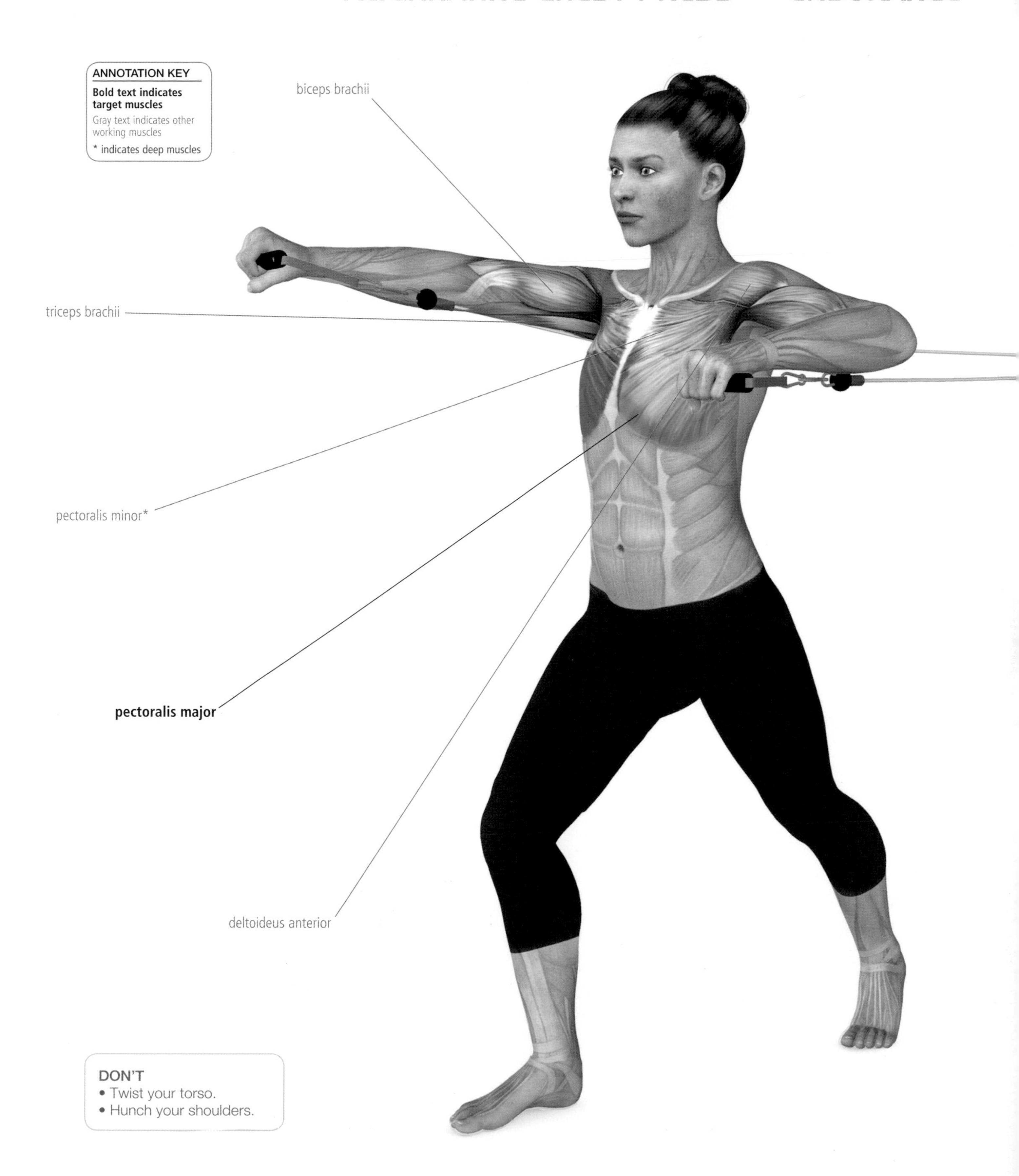

DON'T
- Twist your torso.
- Hunch your shoulders.

ONE-ARMED ROW

ENDURANCE

1. Stand with one leg extended several feet in front of the other, with your front leg bent and your back heel off the floor. Place one end of your resistance band beneath your front foot and grasp the other end with your opposite hand. Rest your free hand above your knee, and lean forward slightly.

a

BEST FOR
- **latissimus dorsi**

2. Bend your arm as you pull the resistance band up toward your chest.

3. Lower and repeat, completing 20 repetitions. Switch sides and repeat.

b

DO
- Keep your back flat throughout the exercise.
- Lean forward so that your back leg and your torso form a straight line.
- Move smoothly and with control, engaging your arm muscles.

TARGETS
- Back

LEVEL
- Intermediate

BENEFITS
- Improves back stabilization
- Strengthens and tones biceps

AVOID IF YOU HAVE . . .
- Lower-back issues

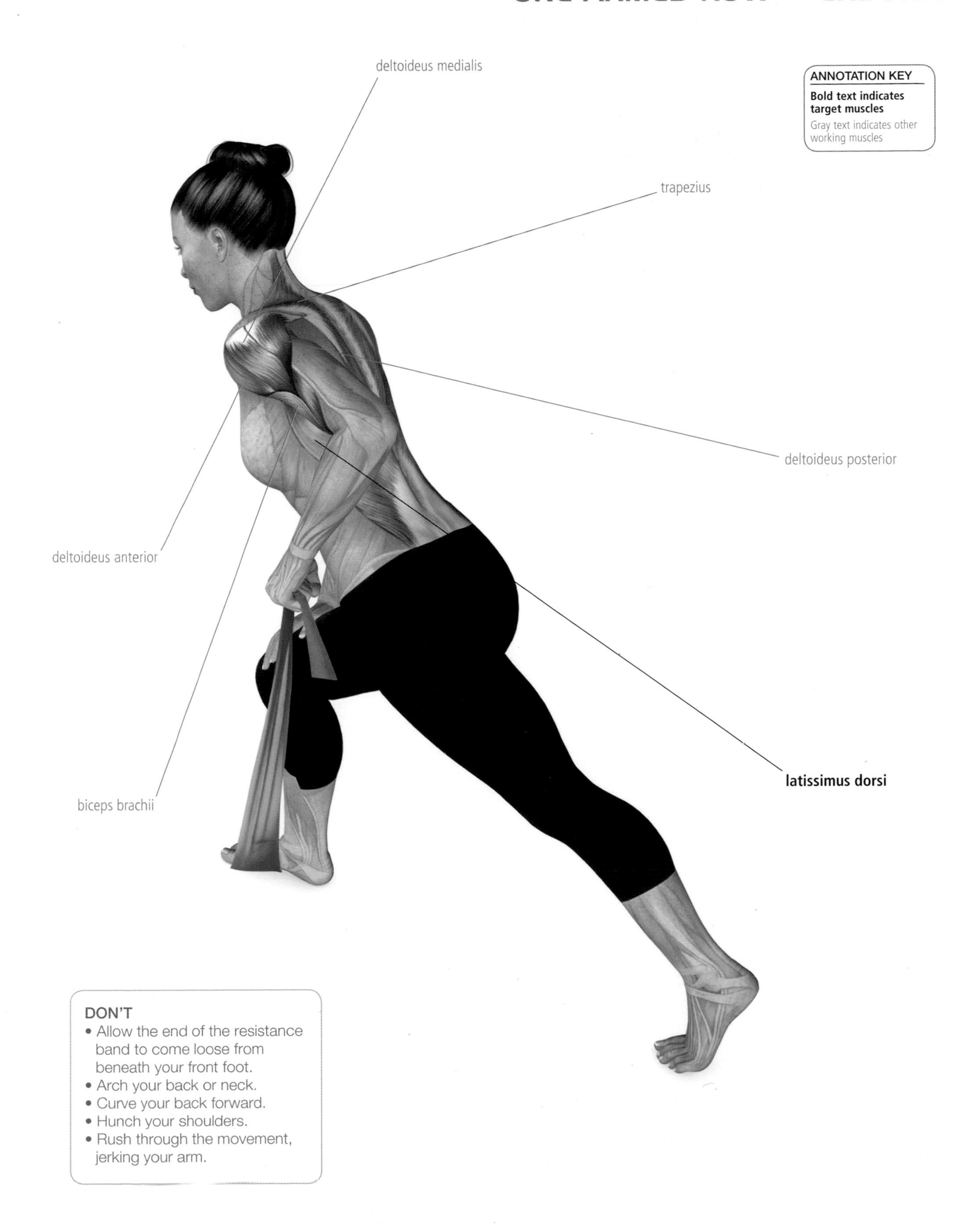

DON'T
- Allow the end of the resistance band to come loose from beneath your front foot.
- Arch your back or neck.
- Curve your back forward.
- Hunch your shoulders.
- Rush through the movement, jerking your arm.

OVERHEAD PRESS

1. Stand upright with one leg extended about a foot behind you, heel off the ground. Position the resistance band beneath the foot of your front leg. Hold the handles in both hands, with arms bent, so that the resistance band is taut.

a

2. Straighten both arms so that they are extended to full lockout above your head a few inches in front of your shoulders.

b

3. Lower your arms to starting position and then repeat. Perform three sets of 15.

c

TARGETS
- Shoulders
- Triceps

LEVEL
- Beginner

BENEFITS
- Strengthens and tones shoulders and upper arms

AVOID IF YOU HAVE . . .
- Shoulder issues

BEST FOR
- deltoideus anterior

ANNOTATION KEY

Bold text indicates target muscles

Gray text indicates other working muscles

* indicates deep muscles

DO

- Keep the rest of your body stable as you extend your arms.
- Gaze forward throughout the exercise.
- Keep your abs engaged and pulled in.
- Extend both arms at the same time.

DON'T

- Twist your torso.

LATERAL RAISE

ENDURANCE

1. Stand upright with your arms at your sides and feet planted hip-width apart on top of your resistance band. Hold one handle in each hand, palms facing inward.

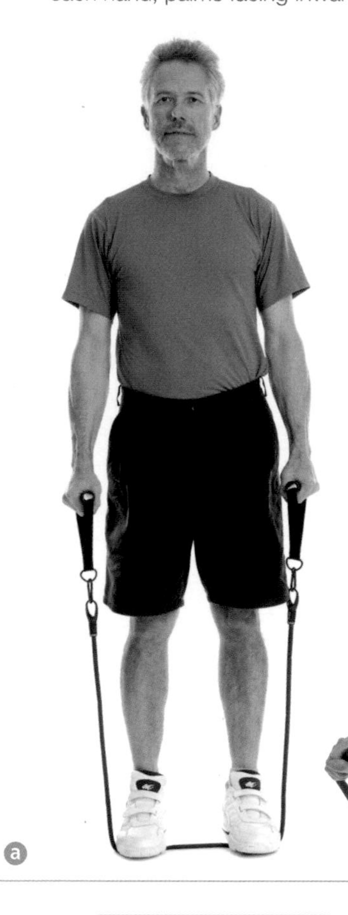

DON'T
- Rush through the movement or jerk your arms.
- Lift your arms above shoulder height.
- Move your feet.

2. Keeping your palms down, raise your arms out to your sides so that they are parallel to the floor.

3. Lower and repeat, completing three sets of 15.

TARGETS
- Shoulder muscles

LEVEL
- Intermediate

BENEFITS
- Strengthens and tones deltoids
- Sculpts triceps

AVOID IF YOU HAVE . . .
- Shoulder issues

BEST FOR
- **deltoideus medialis**

DO
- Raise your arms directly out to the sides.
- Keep the movement slow, smooth, and controlled.
- Keep your torso straight and your gaze forward.

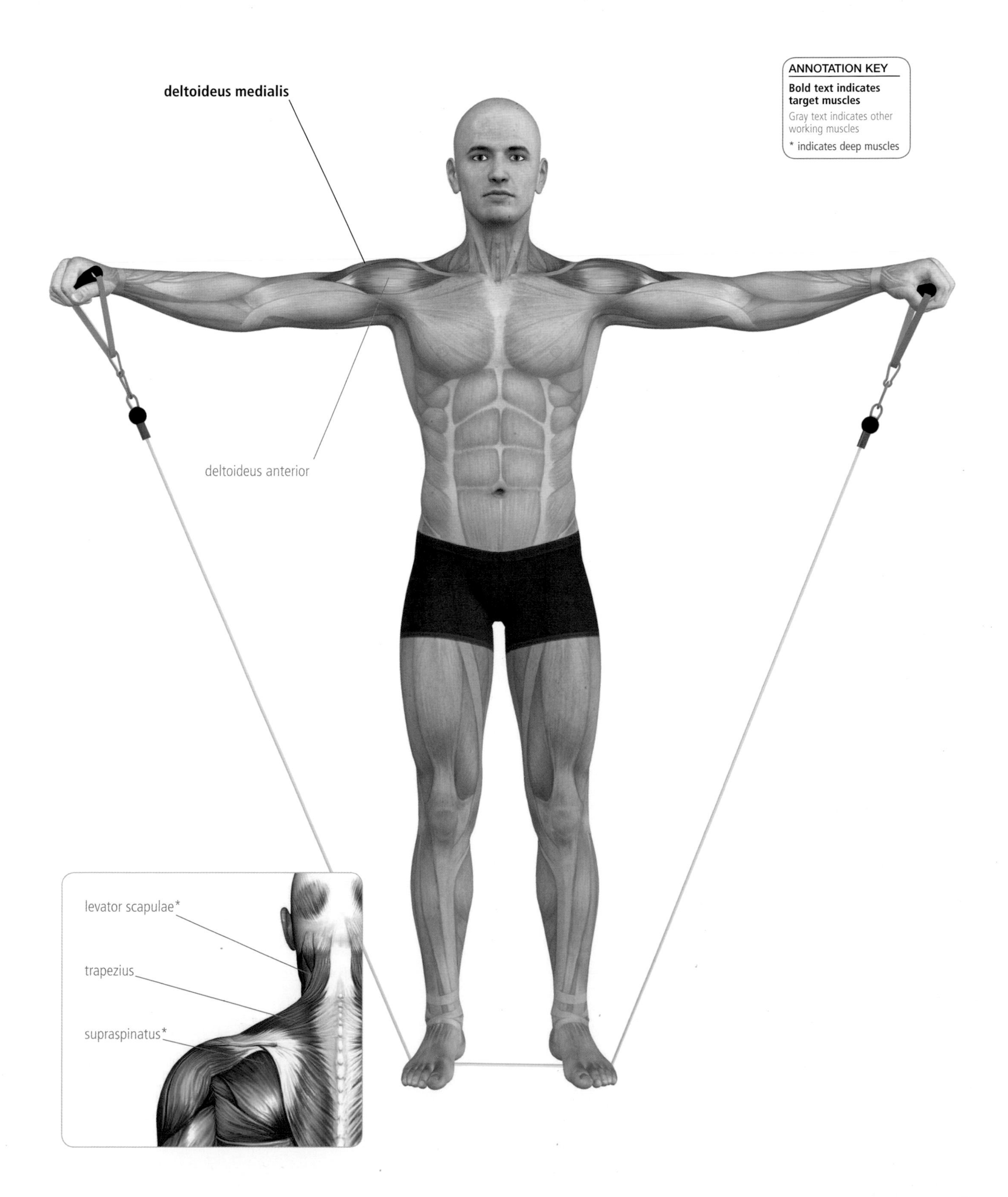
ANNOTATION KEY
Bold text indicates target muscles
Gray text indicates other working muscles
* indicates deep muscles
deltoideus medialis
deltoideus anterior
levator scapulae*
trapezius
supraspinatus*

STANDING FLY

ENDURANCE

❶ Run a resistance band around a sturdy, stable object such as a pole or column. Stand upright, with your feet planted shoulder-width apart and your knees soft. Grasp both of the handles of your resistance band, and extend your arms in front of you to almost shoulder height, holding the band taut.

DON'T
- Arch your back.
- Twist your torso.

BEST FOR
- pectoralis major

TARGETS
- Back
- Chest
- Upper back

LEVEL
- Intermediate

BENEFITS
- Strengthens and tones shoulders and upper back

AVOID IF YOU HAVE . . .
- Lower-back issues
- Shoulder pain

❷ Slowly and with control, bring both arms out to the sides.

❸ Return to starting position and repeat. Complete three sets of 15.

DO
- Keep your arms parallel to the floor.
- Keep your back flat and your torso stable.
- Engage your abs and glutes throughout the exercise.
- Move both arms at the same time.

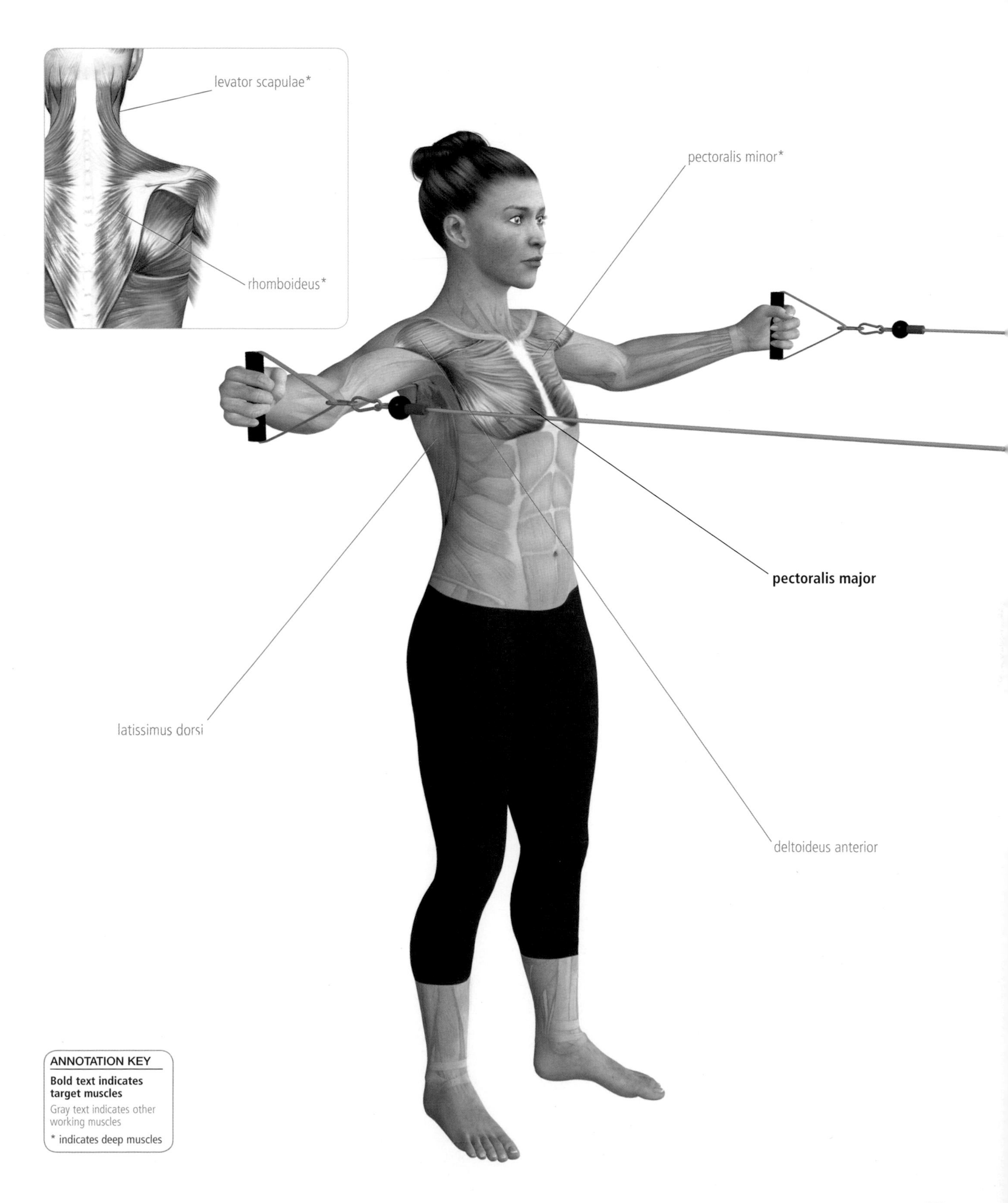

ANNOTATION KEY

Bold text indicates target muscles

Gray text indicates other working muscles

* indicates deep muscles

BICEPS CURL

BEST FOR

- **biceps brachii**

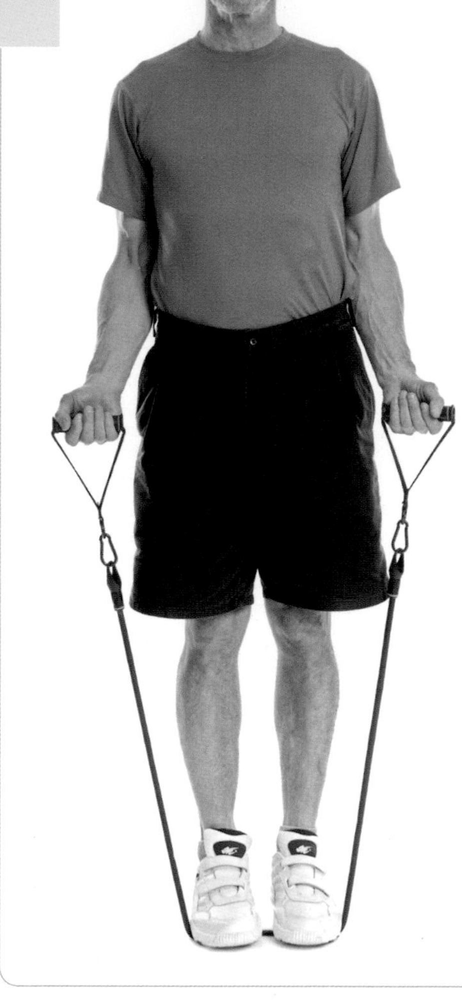

a

1. Stand upright with the resistance band beneath your feet. Your arms should be very slightly bent as you hold both handles of the resistance band in your hands, palms forward.
2. Curl the resistance band upward toward your shoulders.
3. Lower and repeat, completing three sets of 15.

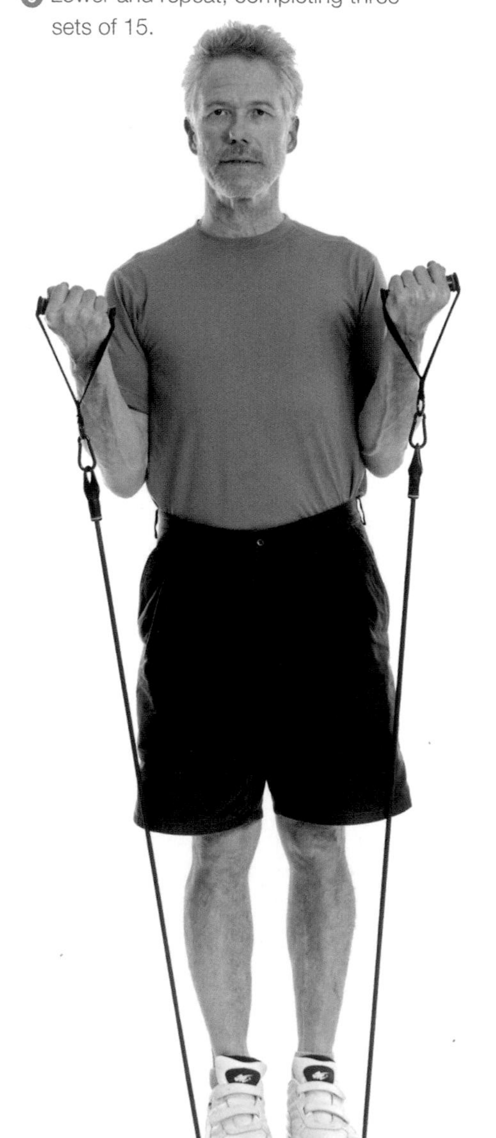

b

TARGETS
- Biceps

LEVEL
- Beginner

BENEFITS
- Strengthens and tones biceps

AVOID IF YOU HAVE . . .
- Wrist or elbow pain

DO
- Keep your elbows at your sides.

DON'T
- Rush through the exercise.

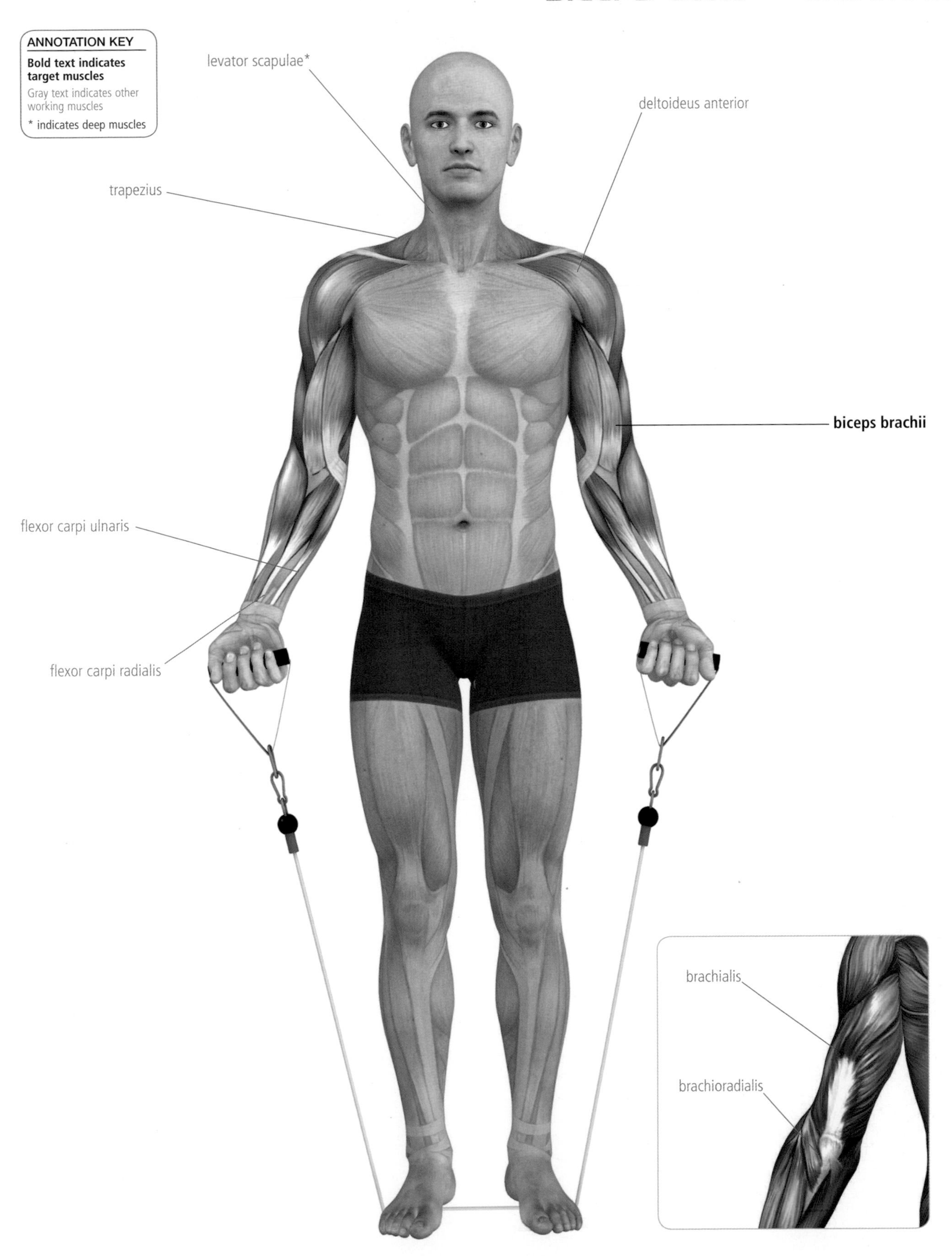
ANNOTATION KEY
Bold text indicates target muscles
Gray text indicates other working muscles
* indicates deep muscles
levator scapulae*
deltoideus anterior
trapezius
biceps brachii
flexor carpi ulnaris
flexor carpi radialis
brachialis
brachioradialis

ONE-ARMED TRICEPS KICKBACK

ENDURANCE

1. Stand in a lunge position, with your front leg bent and your back heel off the ground. Place one end of the resistance band beneath your front foot and grasp the other end in your opposite hand.

2. Lean forward, keeping your back flat so that your torso and back leg form a line. Bend your elbow to position the resistance band next to your hips.

BEST FOR

- triceps brachii

TARGETS
- Triceps

LEVEL
- Intermediate

BENEFITS
- Strengthens and tones triceps

AVOID IF YOU HAVE . . .
- Shoulder issues
- Wrist or elbow pain

DO
- Bend from your hips, not from your waist.
- Try wrapping the resistance band around your front foot.
- Keep your upper arm in place throughout the exercise.
- Rest your free hand above your knee.

3. Keep your upper arm in place as you straighten your arm behind you to full lockout.

4. Lower and repeat, performing 15 repetitions. Switch arms and repeat, working up to three sets of 15 per arm.

DON'T
- Allow the resistance band to come loose from beneath your foot.

SQUAT

ENDURANCE

❶ Stand upright with your feet planted shoulder-width apart and your arms at your sides. Position the resistance band beneath both feet, grasping the handles in both hands.

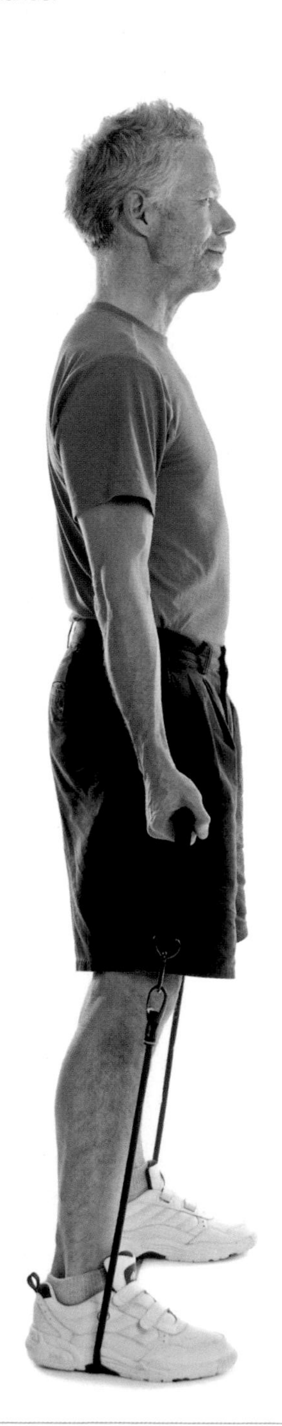

a

❷ Bend your knees to lower your body as you bend your elbows, bringing the resistance band up toward your chest.

❸ Push through your heels as you straighten your legs and bring your arms back to starting position. Repeat, performing three sets of 15.

b

TARGETS
- Buttocks
- Quadriceps

LEVEL
- Intermediate

BENEFITS
- Strengthens and tones core and legs

AVOID IF YOU HAVE . . .
- Lower-back issues

BEST FOR
- **gluteus maximus**
- **rectus femoris**
- **vastus lateralis**
- **vastus intermedius**
- **vastus medialis**

DON'T
- Arch your back or slump forward.
- Twist your torso.
- Lift one arm higher, or at a faster rate, than the other.

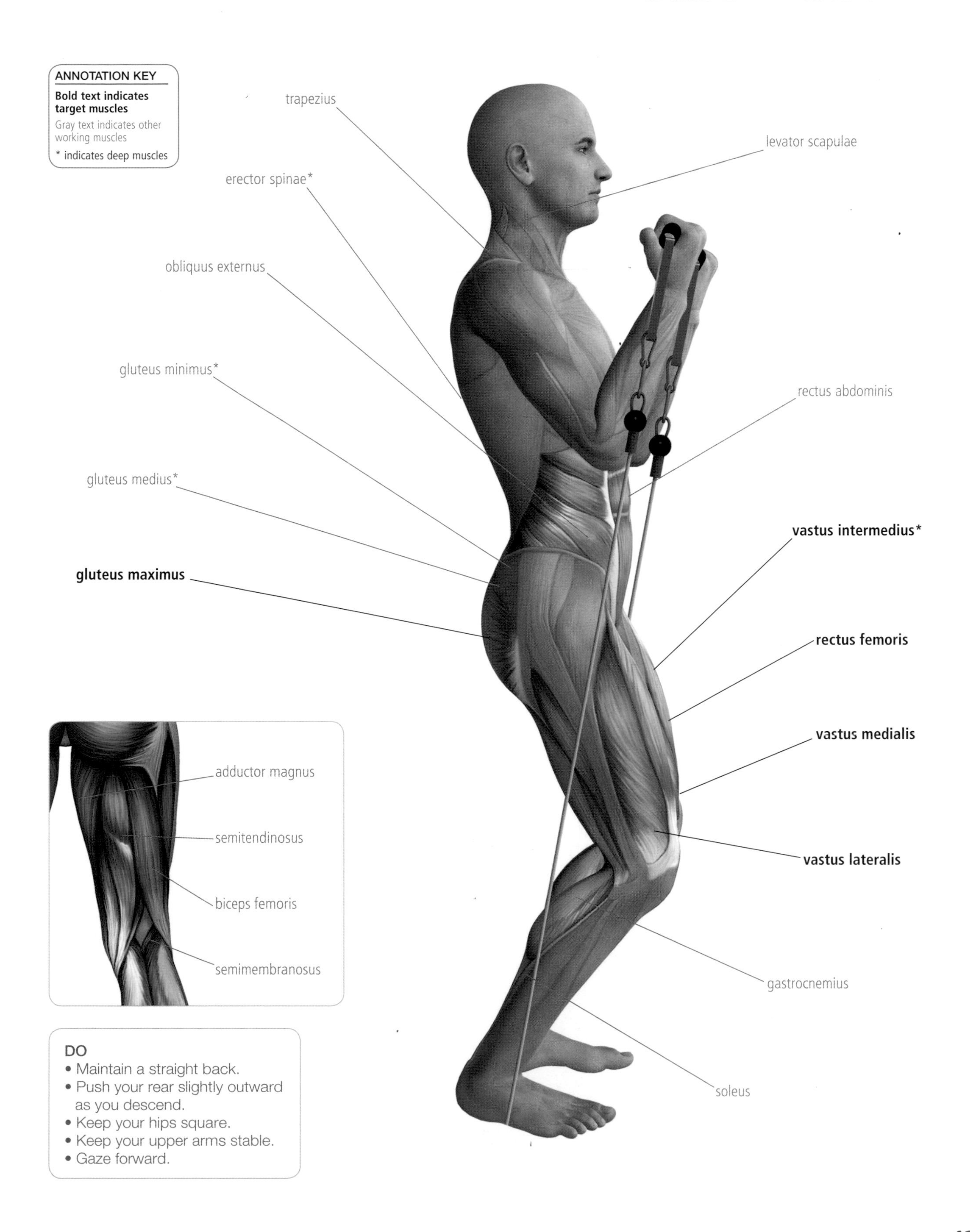

DO
- Maintain a straight back.
- Push your rear slightly outward as you descend.
- Keep your hips square.
- Keep your upper arms stable.
- Gaze forward.

SPLIT SQUAT WITH CURL

ENDURANCE

a

1. Position a resistance band beneath one foot, grasping both handles.
2. Step back so that one leg is several feet behind the other, your front foot anchoring your resistance band and your back heel off the ground. Bend your elbows so that the resistance band is held taut just in front of your torso.

BEST FOR

- gluteus maximus

TARGETS
- Buttocks
- Fronts of thighs

LEVEL
- Intermediate

BENEFITS
- Strengthens and tones glutes and thighs

AVOID IF YOU HAVE . . .
- Knee injury
- Shoulder issues

b

3. Drop your back knee toward the floor, bending both legs until your front thigh is parallel to the ground. At the same time, bring the resistance band closer to your body with palms facing your shoulders.
4. Slowly and with control, straighten your legs as you raise your body, and return your arms to starting position. Complete 15 repetitions, switch sides, and repeat for three sets of 15 on each leg.

DON'T
- Arch your back or allow it to curl forward.
- Twist your torso.
- Let the knee of your back leg touch the ground.
- Hunch your shoulders.

DO
- Keep your back straight and torso upright.
- Gaze forward.

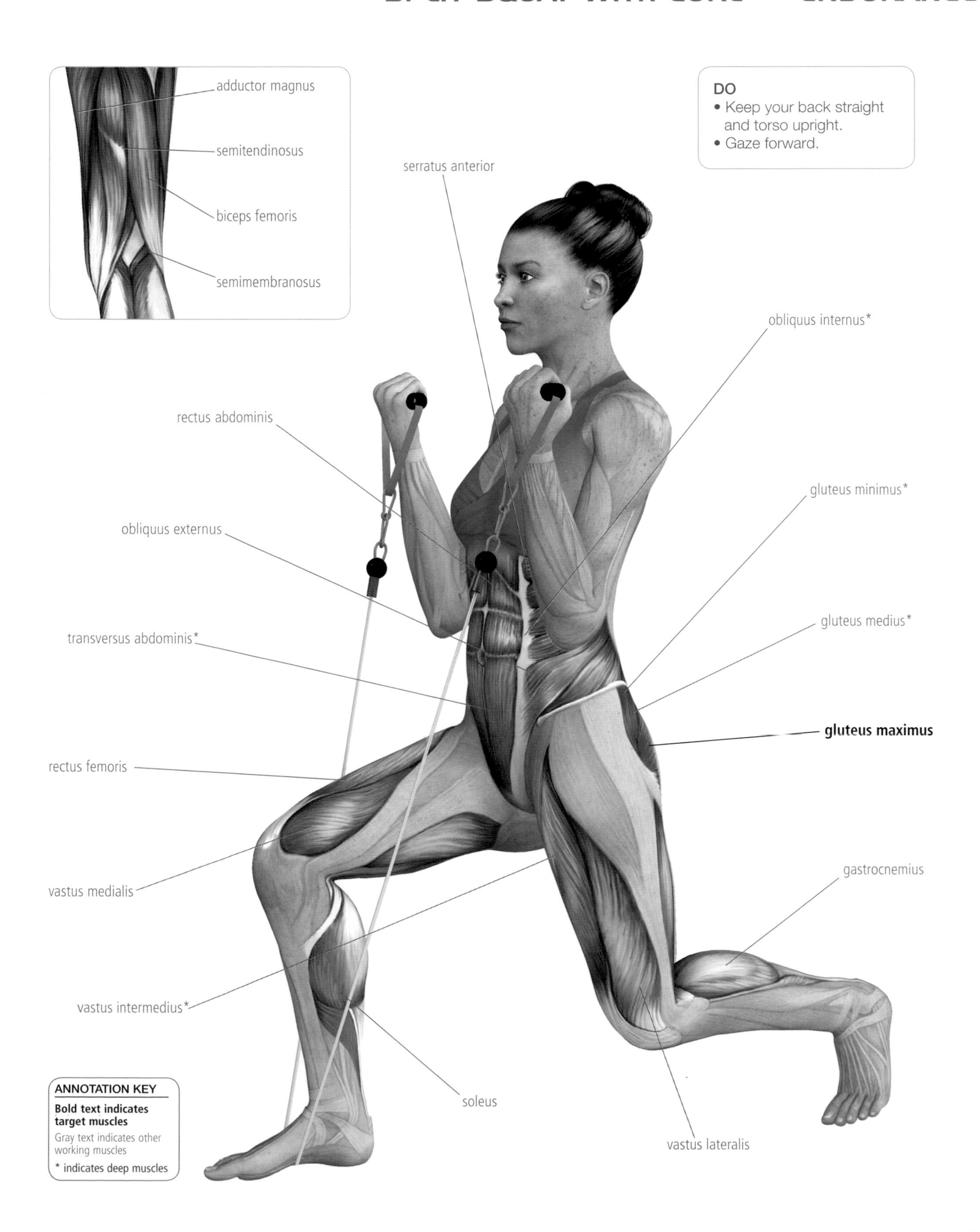

WOOD CHOP WITH RESISTANCE BAND

1. Stand with your feet a little wider than hip-distance apart, the resistance band anchored beneath one of your feet. Hold one handle with both hands, positioning it in front of your body very slightly closer to the anchoring foot.

2. Slowly and smoothly, rotate your core and raise your arms away from the anchoring foot.

3. In a controlled "chopping" motion, return to starting position. Complete 20 repetitions, and then switch sides, performing three sets of 20 per side.

TARGETS
- Obliques

LEVEL
- Beginner

BENEFITS
- Improves core strength and support
- Strengthens and tones obliques

AVOID IF YOU HAVE . . .
- Lower-back issues
- Shoulder issues

BEST FOR
- **obliquus externus**
- **obliquus internus**

DO
- Keep your arms straight.
- Follow your arms with your gaze as you raise and lower your arms.
- Keep your core contracted and your abs engaged.

DON'T
- Twist too jerkily from side to side.
- Raise your arms so high that you lose control of your core and/or arch your back.
- Hunch your shoulders.

ANNOTATION KEY
Bold text indicates target muscles
Gray text indicates other working muscles
* indicates deep muscles

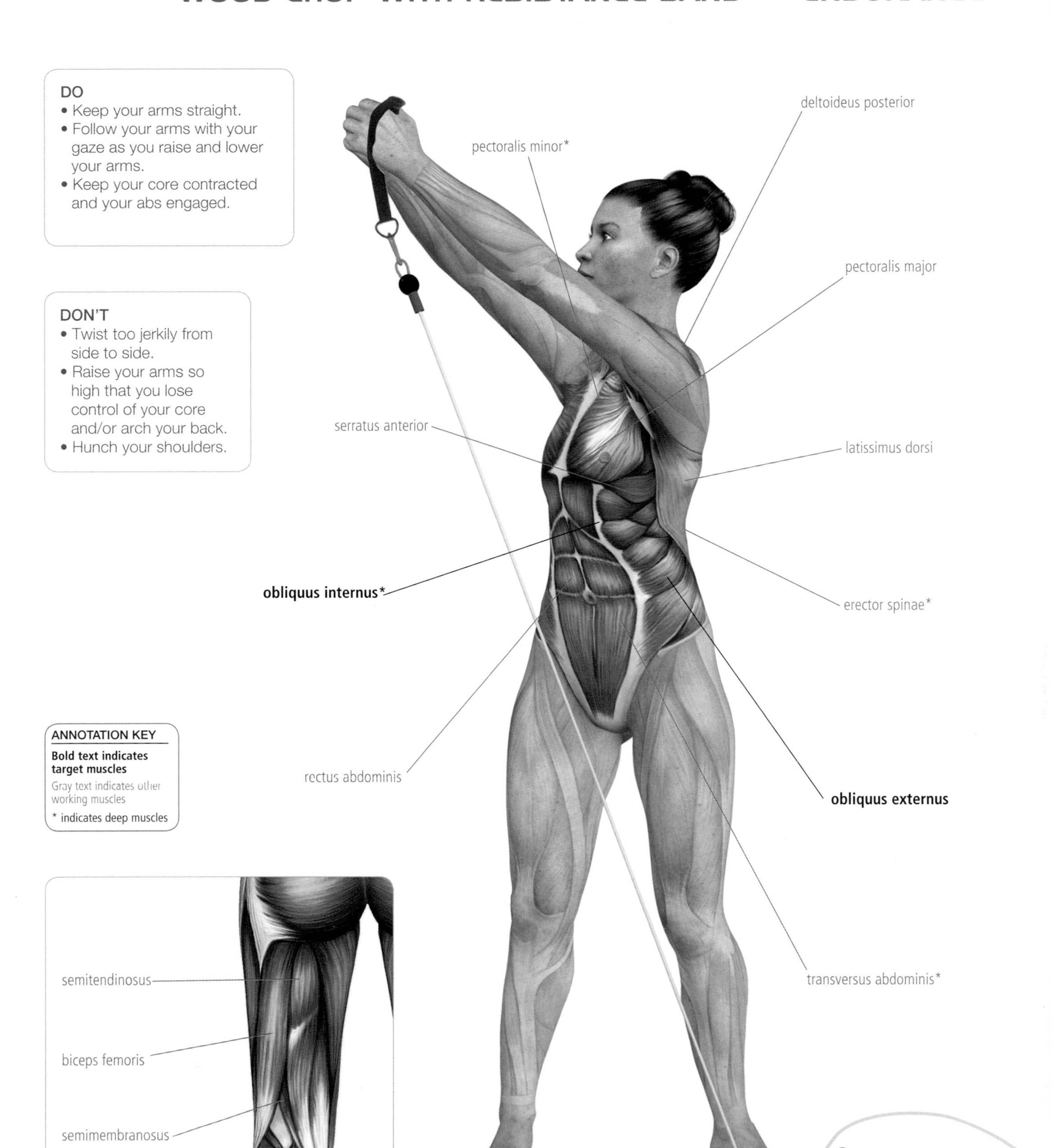

ONE-LEGGED DOWNWARD PRESS

1. Run a resistance band around a sturdy, stable object such as a pole or column. Face the anchoring object, and grasp the handles in both hands. Your arms should be extended straight in front of you at shoulder height, your hands a few inches apart. Bend one knee at a right angle lifting the foot behind you.

2. Keeping one arm stable, lower the other arm to your side.

BEST FOR
- deltoideus anterior

DON'T
- Twist your torso.
- Rush through the exercise.
- Arch your back or neck.

TARGETS
- Abs
- Arms
- Shoulders

LEVEL
- Intermediate

BENEFITS
- Strengthens and tones abs, deltoids, and triceps

AVOID IF YOU HAVE . . .
- Balancing difficulties
- Shoulder issues

MODIFICATION
Easier: Rest the toe of your back foot on the floor.

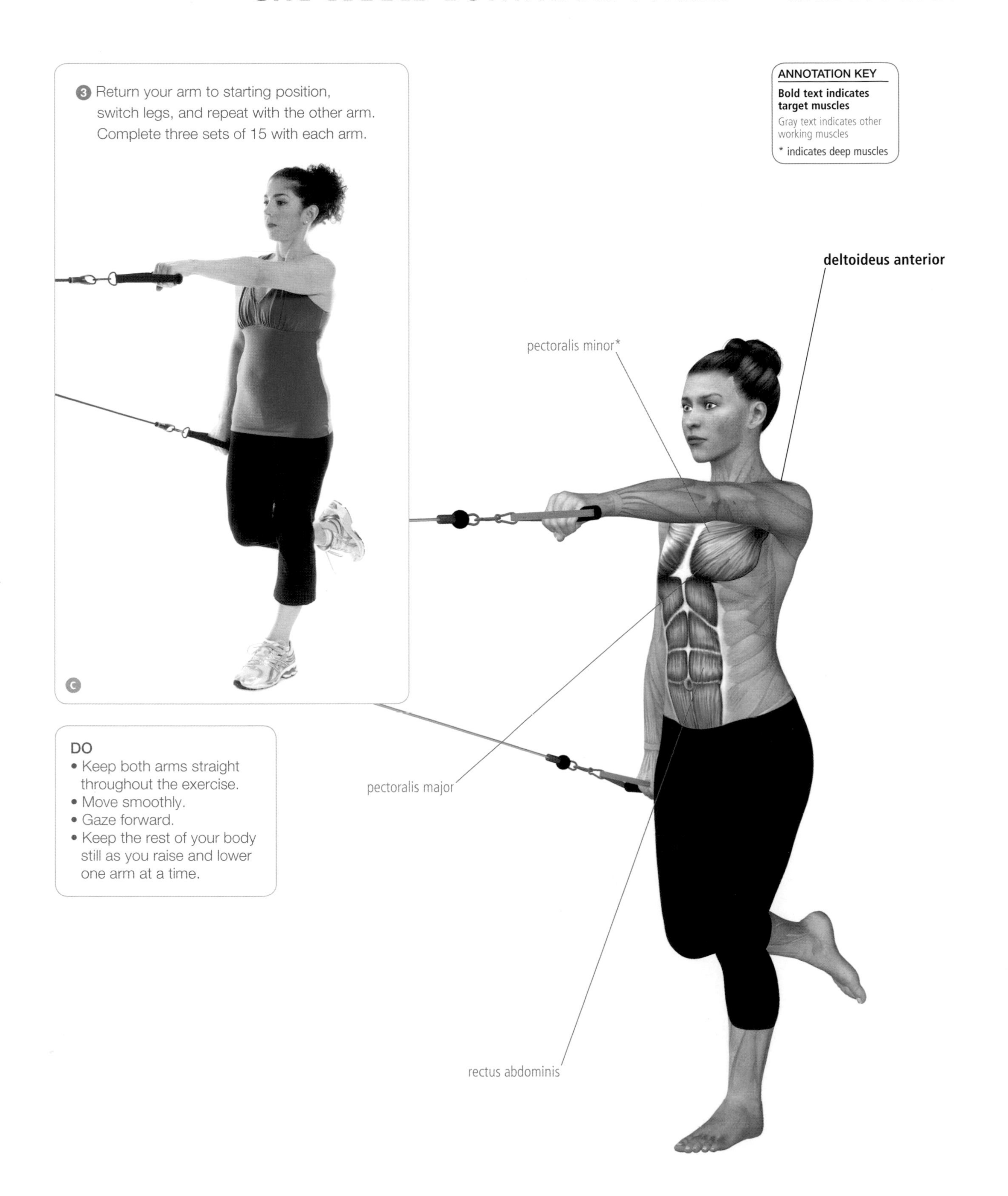

3 Return your arm to starting position, switch legs, and repeat with the other arm. Complete three sets of 15 with each arm.

DO
- Keep both arms straight throughout the exercise.
- Move smoothly.
- Gaze forward.
- Keep the rest of your body still as you raise and lower one arm at a time.

ALIGNMENT & POSTURE EXERCISES

If you are like most people these days, you probably sit behind a steering wheel getting to a job that finds you hunched over a computer keyboard gazing up at a monitor. In the process, you compress your neck and back, strain your shoulders, stress one hip more than the other—and the list goes on. When bones and joints are pushed out of alignment, your posture suffers and your body feels off-kilter. These exercises will help you to feel more balanced, standing straighter and walking taller.

STANDING ABDOMINAL BRACE

1. Stand with your feet planted about hip-width apart and your knees slightly bent. Fold your arms.

2. Bend your legs farther and slightly contract your pelvis. Hold for 5 seconds.

3. Draw your pelvis forward, straighten your legs slightly, and return to starting position. Complete three repetitions.

TARGETS
- Abs
- Lower back

LEVEL
- Beginner

BENEFITS
- Improves posture
- Relieves mild-to-moderate lower-back pain

AVOID IF YOU HAVE . . .
- Severe lower-back pain

ANNOTATION KEY
Bold text indicates target muscles
* indicates deep muscles

rectus abdominis

transversus abdominis*

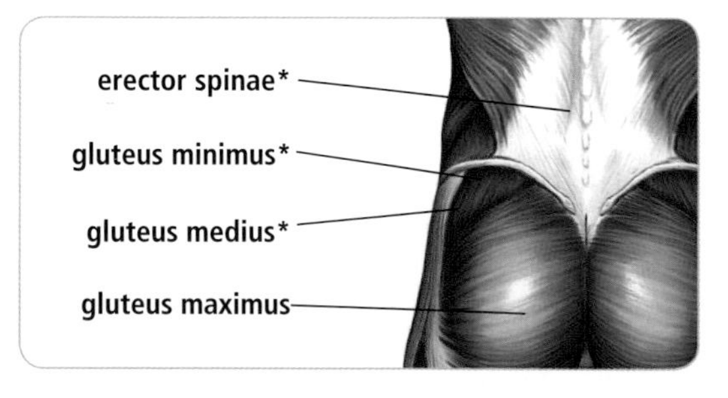

BEST FOR

- **rectus abdominis**
- **transversus abdominis**
- **gluteus maximus**
- **gluteus minimus**
- **gluteus medius**
- **erector spinae**

DON'T
- Rush through the movement.

DO
- Exhale as you contract your core.
- Keep your gaze forward.

LYING PELVIC TILT

BEST FOR

- rectus abdominis
- gluteus maximus
- gluteus minimus
- gluteus medius
- transversus abdominis
- ligamentum interspinalis
- ligamentum longitudinale posterius
- articulationes zygapophysiales
- ligamentum capsular facet

1. Lie on your back with your legs bent and arms extended out to your sides. Your back should be in neutral position, meaning the natural curve of the lumbar spine will slightly elevate your lower back.

2. Contract your abs, and gently pull your navel in toward your spine while tilting your hips upward to flatten your lower back on the floor. Hold for 5 seconds

3. Release and repeat, completing five repetitions.

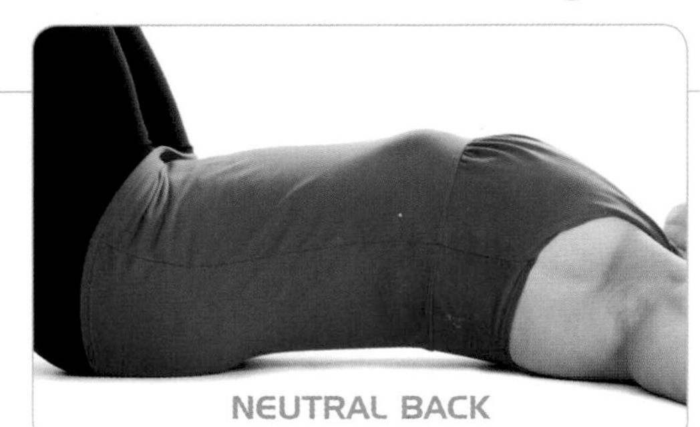

b

rectus abdominis

transversus abdominis*

iliopsoas*

pectineus*

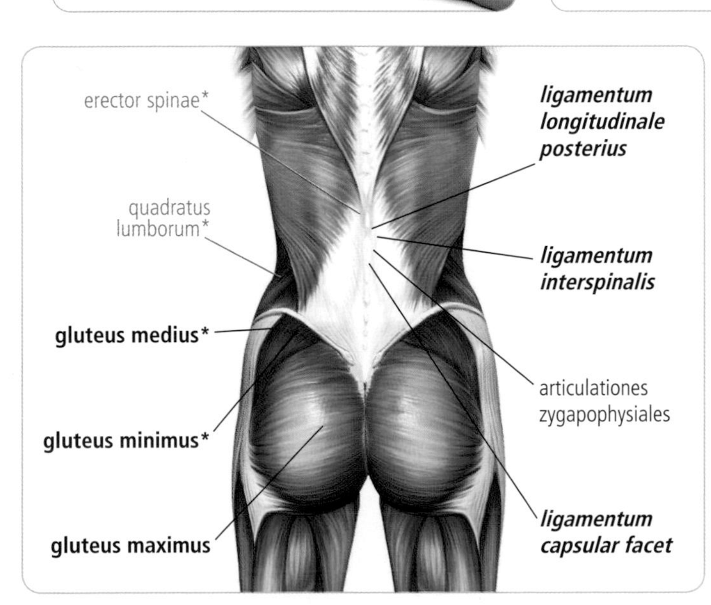

ANNOTATION KEY

Bold text indicates target muscles

Gray text indicates other working muscles

Black text indicates joints

Italic text indicates ligaments

* indicates deep muscles

DON'T

- Rush through this exercise.
- Let your feet or glutes come off the floor.

DO

- Keep the rest of your body stationary as you move your pelvis.

TARGETS

- Abs
- Lower back

LEVEL

- Beginner

BENEFITS

- Improves posture
- Relieves mild-to-moderate lower-back pain

AVOID IF YOU HAVE . . .

- Severe lower-back pain

BRIDGE

ALIGNMENT & POSTURE

❶ Lie on your back with your legs bent, your feet flat on the floor, and your arms extended at your sides, angled slightly away from the body.

❷ Push through your heels while raising your glutes off the floor. With your feet and thighs parallel, push your arms into the floor, while extending through your fingertips.

❸ Lengthen your neck away from your shoulders as you lift you hips higher so that you form a straight line from shoulder to knee.

❹ Hold for 30 seconds to 1 minute. Exhale to release your spine slowly to the floor. Repeat three times.

TARGETS
- Glutes
- Hamstrings
- Quadriceps
- Lower back
- Hips

LEVEL
- Beginner

BENEFITS
- Strengthens glutes, quadriceps, and hamstrings
- Stabilizes core

AVOID IF YOU HAVE . . .
- Shoulder injury
- Neck injury
- Back injury

DON'T
- Overextend your abdominals past your thighs in the finished position.
- Arch your back.

DO
- Push through your heels, not your toes.
- Keep your knees and feet aligned.
- Keep your arms and feet on the floor.

ANNOTATION KEY

Bold text indicates target muscles

Gray text indicates other working muscles

* indicates deep muscles

BEST FOR

- erector spinae
- iliopsoas
- sartorius
- rectus femoris
- gluteus maximus
- gluteus medius
- gluteus minimus
- vastus lateralis
- vastus intermedius
- vastus medialis

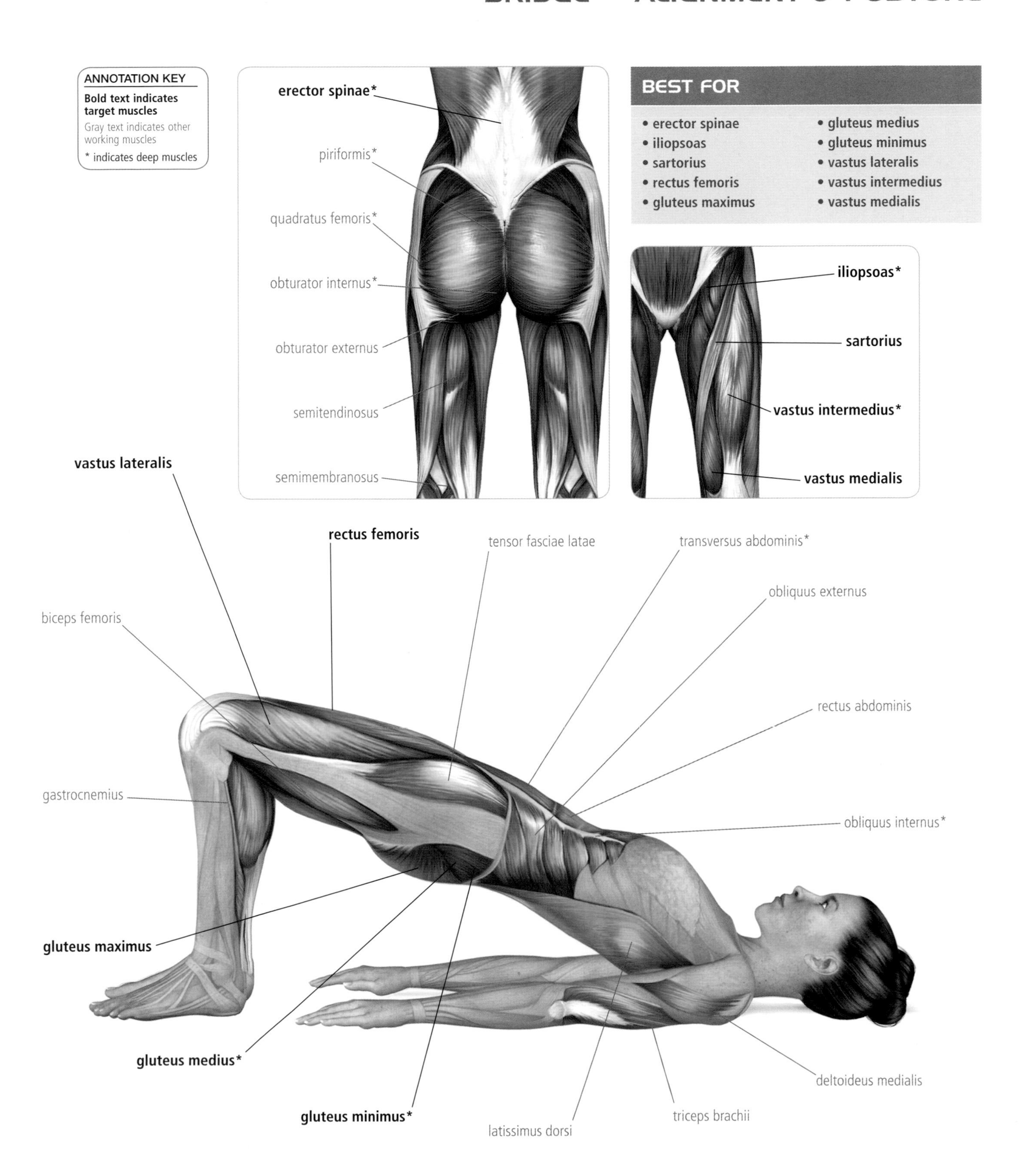

PRONE LIMB RAISES

ARM RAISE

1. Lie facedown with your arms and legs extended so that your body forms a straight line. Keep your arms and legs relaxed.

a

BEST FOR

- **erector spinae**
- **latissimus dorsi**
- **deltoideus posterior**

2. Raise one arm up while keeping your upper back as still as possible.

b

3. Lower, and repeat with the opposite arm, performing five repetitions per arm.

c

TARGETS
- Lower back
- Shoulders

LEVEL
- Beginner

BENEFITS
- Aligns spine
- Maintains spinal range of motion
- Stretches the back

AVOID IF YOU HAVE . . .
- Shoulder issues

trapezius
infraspinatus*
deltoideus posterior
supraspinatus*
teres minor
subscapularis*
latissimus dorsi
erector spinae*

ANNOTATION KEY
Bold text indicates target muscles
Gray text indicates other working muscles
* indicates deep muscles

DO
- Keep your legs still.
- Perform the raises slowly and with control; each repetition should take about five seconds.
- Gaze downward, keeping your neck long.

DON'T
- Arch your back or neck.
- Move your torso as you raise or lower your arm.
- Rush through the movement.

BEST FOR
- **iliopsoas**
- **erector spinae**

❶ Lie facedown with your legs extended. Cross your arms and rest your chin on them for stability.

❷ Keeping your upper body still and your legs straight, raise one leg off the ground while keeping your knees locked.

❸ Flex your foot, lower your leg, and repeat with the other leg for three repetitions of 10-second counts per leg.

DON'T
- Sacrifice form for speed.

ANNOTATION KEY
Bold text indicates target muscles
Gray text indicates other working muscles
* indicates deep muscles

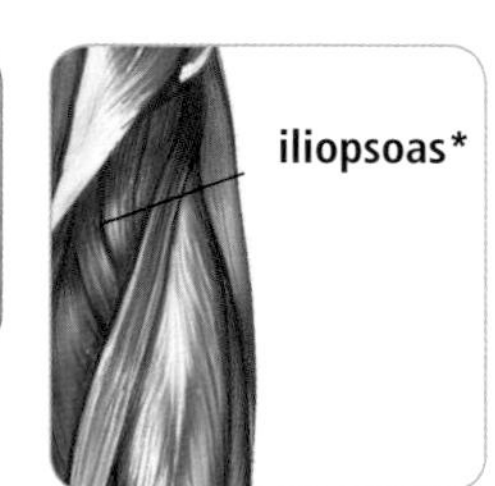

DO
- Keep your legs straight as you raise and lower them.
- Engage your glutes.

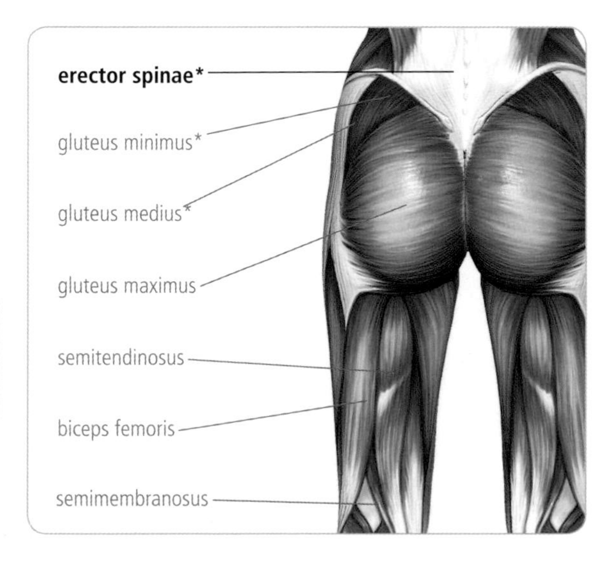

TARGETS
- Lower-back
- Hip flexors

LEVEL
- Beginner

BENEFITS
- Aligns spine
- Maintains spinal range of motion
- Stretches the back

AVOID IF YOU HAVE . . .
- Severe lower-back pain

SWISS BALL PELVIC TILT

1. Sit upright on your Swiss ball, with your feet flat on the floor and hands resting on your knees or thighs.

BEST FOR
- rectus abdominis
- transversus abdominis
- gluteus maximus
- gluteus minimus
- gluteus medius
- erector spinae

TARGETS
- Lower back
- Abs
- Glutes

LEVEL
- Beginner

BENEFITS
- Improves posture
- Relieves mild-to-moderate lower-back pain

AVOID IF YOU HAVE . . .
- Severe lower-back pain

DON'T
- Rush through the movement.

2. Tilt your pelvis forward, using the motion of the ball to assist you. Contract your abdominals, and hold for 5 seconds.

3. Return to starting position, and contract again. Repeat the back-and-forth motion, holding each position for 5 seconds for 10 reps.

DO
- Position your hips over the center of the ball so that you are fully supported.
- Exhale while contracting.

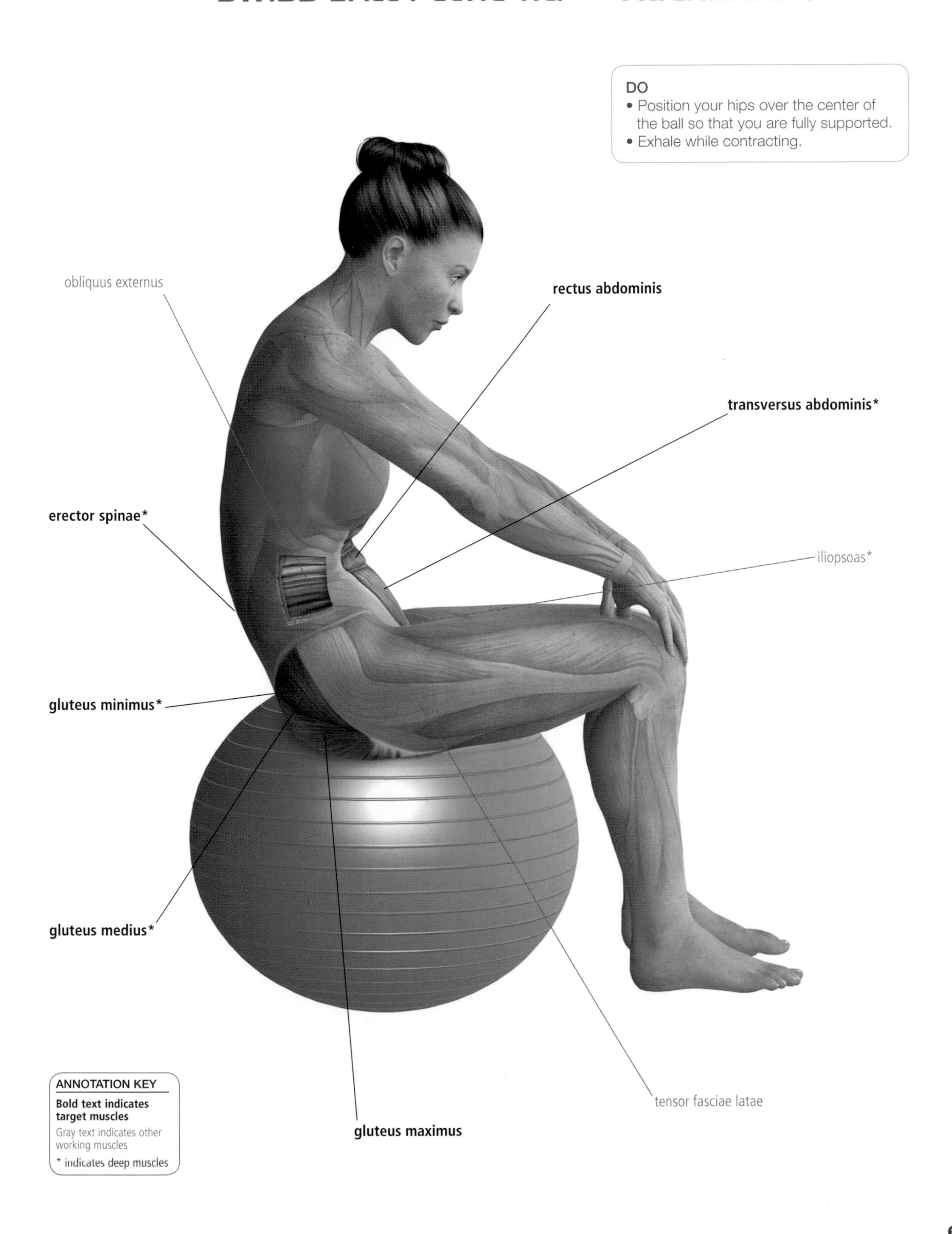

ANNOTATION KEY
Bold text indicates target muscles
Gray text indicates other working muscles
* indicates deep muscles

SWISS BALL SIT-TO-BRIDGE

BEST FOR

- deltoideus medialis
- iliopsoas
- latissimus dorsi
- serratus anterior
- pectoralis major
- pectoralis minor
- ligamentum longitudinale anterius
- gluteus maximus
- gluteus medius
- gluteus minimus
- tensor fasciae latae
- quadratus lumborum
- quadratus femoris

1. Sit upright on your Swiss ball, with your feet flat on the floor and hands resting on your knees or thighs.
2. Extend your arms in front of you, and slowly step forward while leaning back on the ball, allowing it to roll up your spine.

3. As you extend your arms back and over your head, walk your feet forward so that the ball continues to roll up your spine.

4. Roll back until your hands are on the floor, with your arms slightly bent. Hold this bridged position for 5 seconds, ending in an exhale.
5. To release the stretch, bend your knees, from your hips to the floor, lift your head from the ball, and then walk back to the starting position.

TARGETS
- Upper and middle back
- Abs

LEVEL
- Advanced

BENEFITS
- Stretches upper back
- Increases spinal extension
- Stretches abs

AVOID IF YOU HAVE . . .
- Lower-back issues
- Balancing difficulties

DO
- Position your hips over the center of the ball so that you are fully supported.
- Lean back slowly and with control.
- Exhale while contracting.
- Tuck your chin slightly as you begin to roll backward.

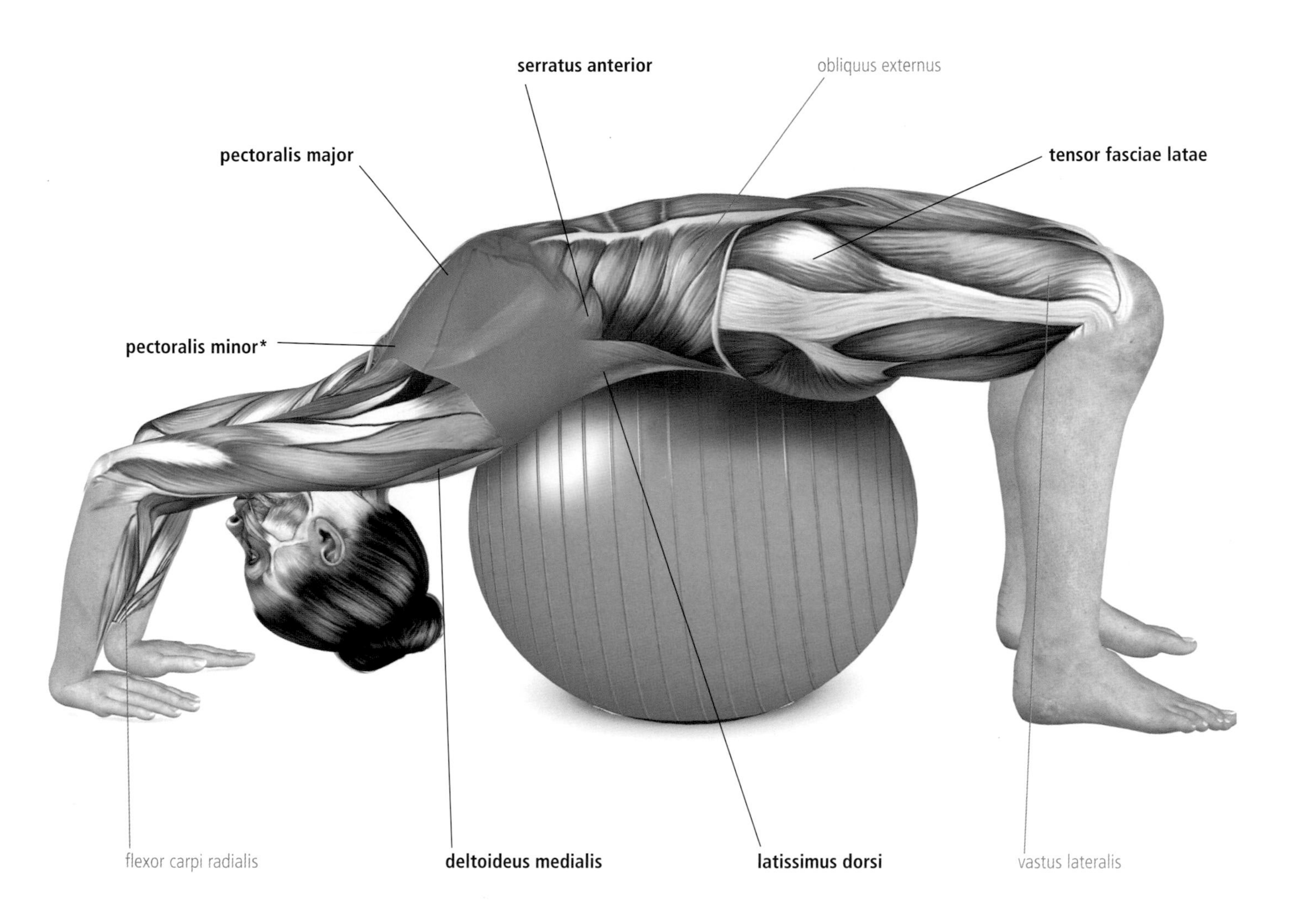

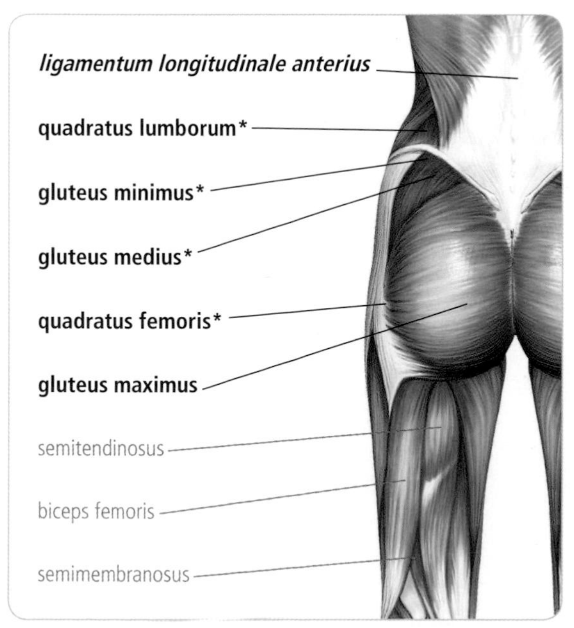

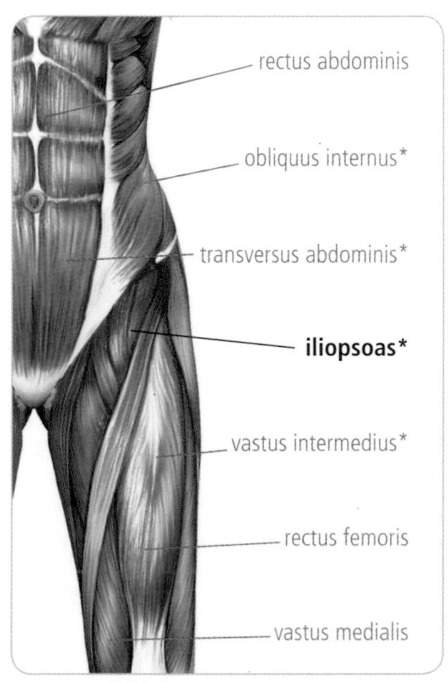

DON'T
- Roll too quickly.
- Allow your feet to come off the floor.
- Allow the ball to shift laterally.

ANNOTATION KEY

Bold text indicates target muscles

Gray text indicates other working muscles

Italic text indicates ligaments

* indicates deep muscles

COBRA STRETCH

1. Lie supine on the floor, with your legs extended behind you, slightly apart. Gaze straight ahead, and bend your arms so that your forearms and palms rest flat on the floor.

BEST FOR

- erector spinae
- quadratus lumborum
- latissimus dorsi
- gluteus maximus
- gluteus medius
- pectoralis major
- rectus abdominis
- deltoideus medialis
- teres major
- teres minor

3. Press downward with your palms, and slowly lift through the top of your chest as you straighten your arms until they are fully extended.

4. Pull your tailbone down toward your pubis as you push your shoulders down and back. Elongate your neck and gaze forward.

5. Hold for up to 15 seconds. Release and repeat, performing three repetitions.

TARGETS
- Spinal joints

LEVEL
- Intermediate

BENEFITS
- Strengthens spine
- Stretches, chest, abs, and shoulders

AVOID IF YOU HAVE . . .
- Lower-back injury

DON'T
- Tip your head back.
- Jerk your body as you lift; instead, move slowly and with control.
- Twist your torso.
- Overdo the stretch—arching too far backward can place excessive pressure on your lower back.

ANNOTATION KEY
Bold text indicates target muscles
Gray text indicates other working muscles
* indicates deep muscles

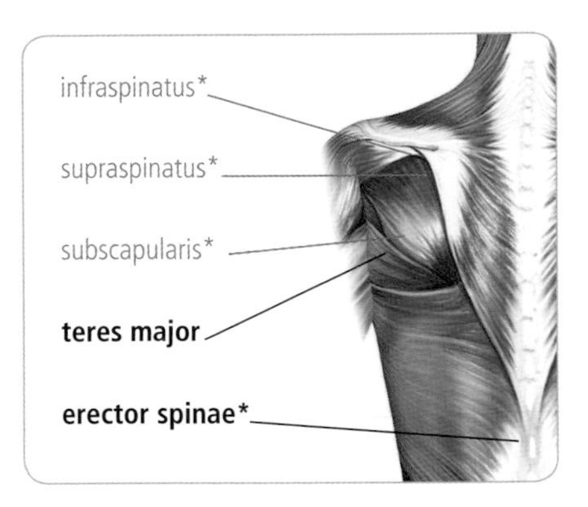

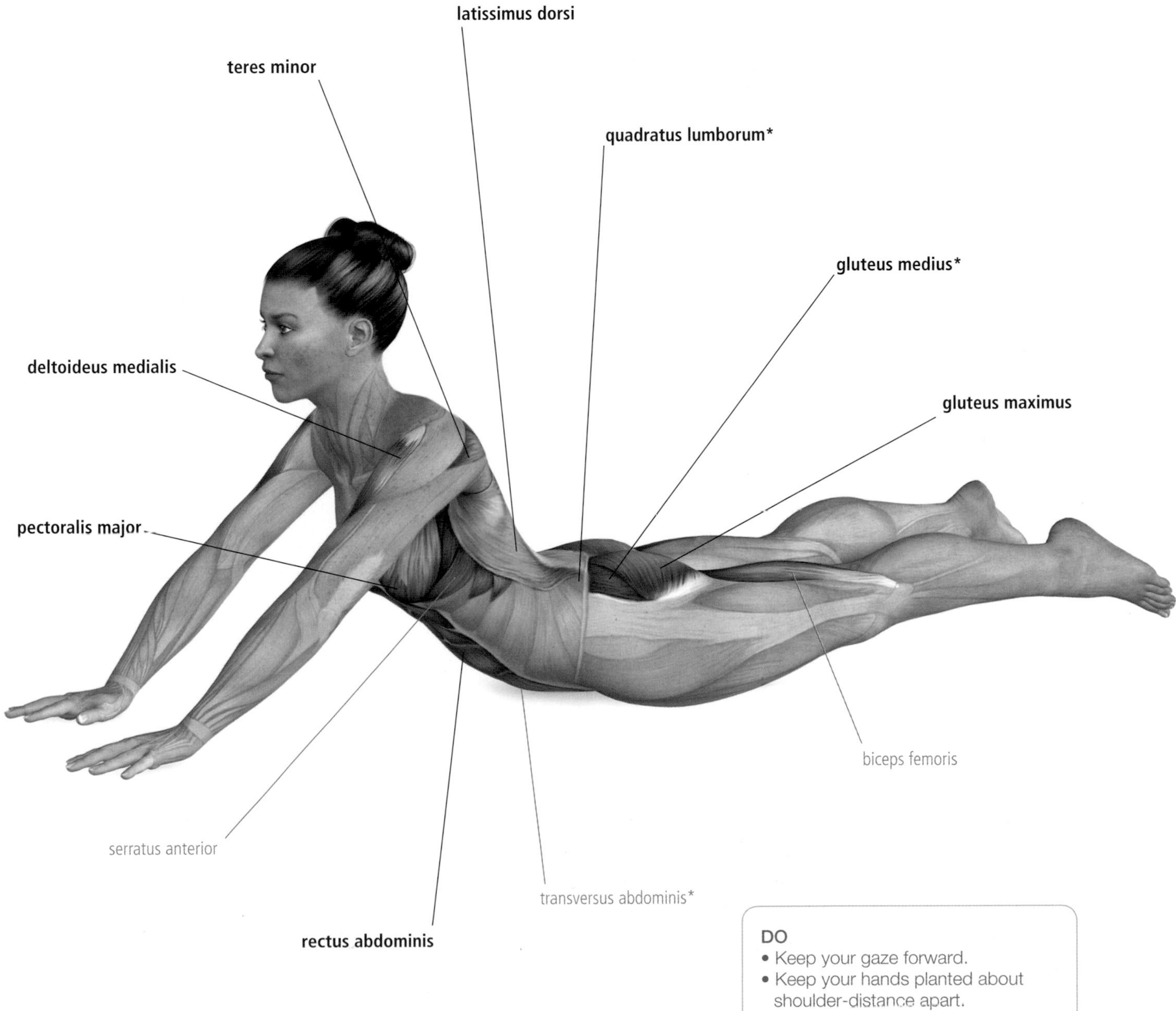

DO
- Keep your gaze forward.
- Keep your hands planted about shoulder-distance apart.

BILATERAL SCAPULAR RETRACTION

SWISS BALL ARM FLEXION

1. Kneel on the floor with your forehead resting on your Swiss ball.
2. Extend your arms in front of the ball, and raise them slightly. Squeeze your shoulder blades together. Hold for 5 seconds.
3. Release your arms and then repeat, completing four repetitions.

TARGETS
- Shoulders
- Upper back

LEVEL
- Intermediate

BENEFITS
- Stabilizes shoulder joints
- Aids spinal alignment

AVOID IF YOU HAVE . . .
- Shoulder injury
- Neck pain
- Knee issues

DO
- Keep your abs tight and engaged.

ANNOTATION KEY
Bold text indicates target muscles
Gray text indicates other working muscles
* indicates deep muscles

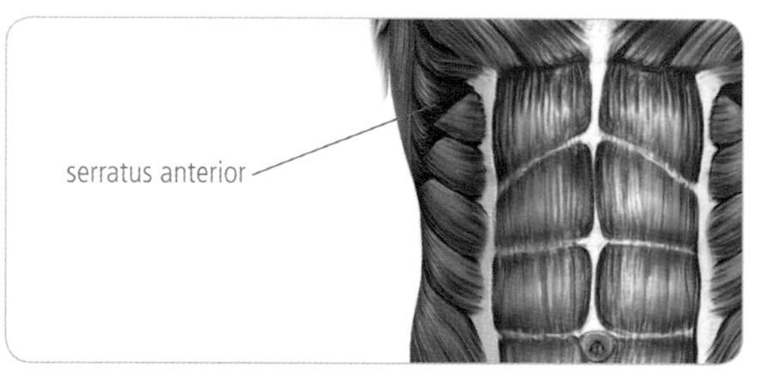

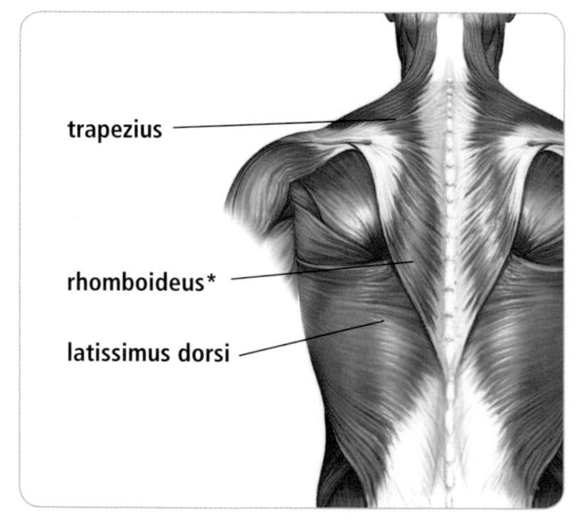

BEST FOR
- **rhomboideus**
- **trapezius**
- **latissimus dorsi**

DON'T
- Allow your abs to relax.
- Allow the ball to move; only your arms should be moving, while your forehead should stay in place on the ball.

SWISS BALL ARM EXTENSION

1. Kneel on the ground with your forehead resting on the Swiss ball.
2. Keeping your abs tight, extend your arms behind you and raise them slightly. Squeeze your shoulder blades together. Hold for 5 seconds.
3. Release your arms, and then repeat, completing four repetitions.

BEST FOR

- **rhomboideus**
- **trapezius**
- **latissimus dorsi**

serratus anterior

TARGETS
- Shoulders
- Upper back

LEVEL
- Intermediate

BENEFITS
- Stabilizes shoulder joints
- Aids spinal alignment

AVOID IF YOU HAVE . . .
- Shoulder injury
- Neck pain
- Knee issues

DON'T
- Twist your torso.
- Arch your neck.

DO
- Keep your arms elevated.
- Keep your legs in place as you stretch.

trapezius

rhomboideus*

latissimus dorsi

ANNOTATION KEY

Bold text indicates target muscles

Gray text indicates other working muscles

* indicates deep muscles

RANGE OF MOTION EXERCISES

Keeping your muscles and joints healthy involves tuning in to their range of motion. The following exercises focus on the specific parts of the body that you need to keep limber: the neck, shoulders, elbows, wrists, knees, hips, and ankles. It's beneficial to target areas in which you notice compromised range of motion, but it's also important to maintain your body's full range of motion in areas that are working well. Practiced consistently, these exercises improve blood flow and bolster flexibility.

NECK & SHOULDER EXERCISES

HEAD TURN

❶ Sit or stand upright, gazing forward with your neck straight.

❷ Turn your head to one side to look over your shoulder. Hold for 10 seconds

❸ Face forward, then repeat on the other side. Complete 10 reps on each side.

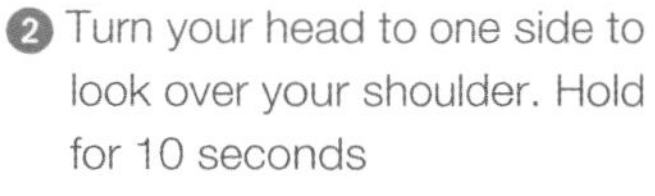

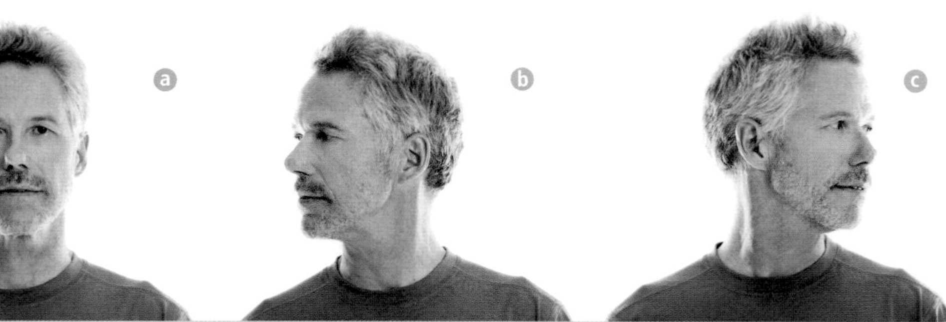

BEST FOR

- **sternocleidomastoideus**
- **trapezius**
- **ligamentum nuche**
- **ligamentum supraspinous**

DO
- Keep your shoulders down and relaxed.

DON'T
- Turn your head too far to the side.

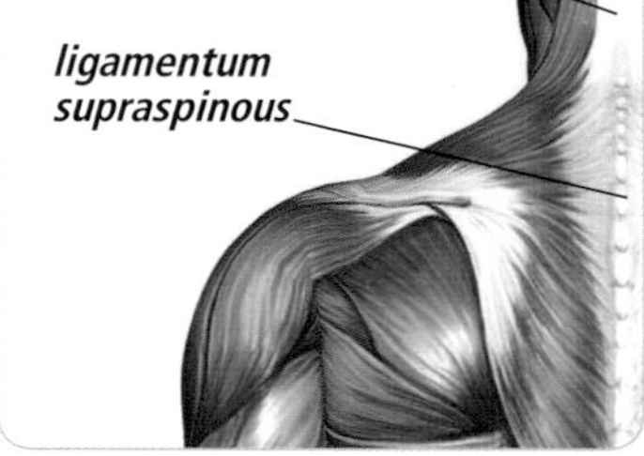

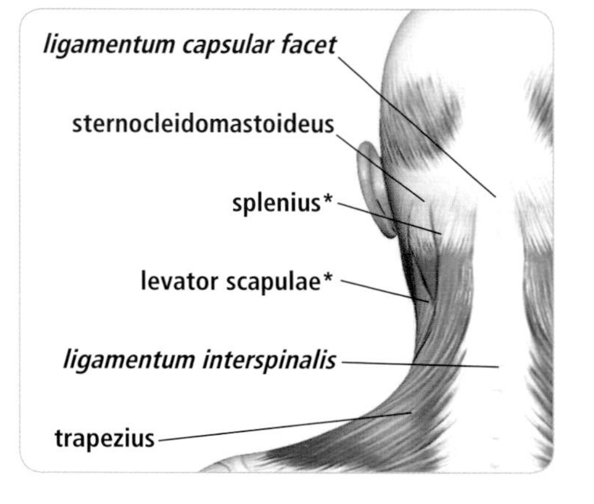

TARGETS
- Neck muscles

LEVEL
- Beginner

BENEFITS
- Improves neck's range of motion
- Relieves neck pain

AVOID IF YOU HAVE . . .
- Numbness running down your arm or into your hand

ANNOTATION KEY

Bold text indicates target muscles

Gray text indicates other working muscles

Italic text indicates ligaments

* indicates deep muscles

DON'T
- Tilt your head so far up or down that you feel strain in your neck muscles.

HEAD TILT

❶ Sit or stand upright, gazing forward with your neck straight.

❷ Gently tilt your head back to look up at the ceiling. Hold for 10 seconds.

❸ Release the stretch, and pull your chin forward to look at the floor. Hold for 10 seconds.

❸ Repeat entire sequence, tilting up and then down 10 times.

BEST FOR

- **splenius**
- **sternocleidomastoideus**
- **levator scapulae**
- **trapezius**
- **ligamentum interspinalis**
- **ligamentum capsular facet**

BACK PAT & RUB

1. Stand, keeping your neck, shoulders, and torso straight.
2. Bring one arm up toward the ceiling, and bend the elbow, bringing your hand down the center of your back.
3. Bring your other arm out to the side, bend the elbow, and bring your arm up the center of the back.
4. Hook your fingers together behind your back, and draw both elbows toward the center. Hold for 30 seconds to 1 minute. release your arms, and repeat on the other side.

a

b

DON'T
- Strain—if your hands cannot meet, simply slide them as close to each other as you can reach.

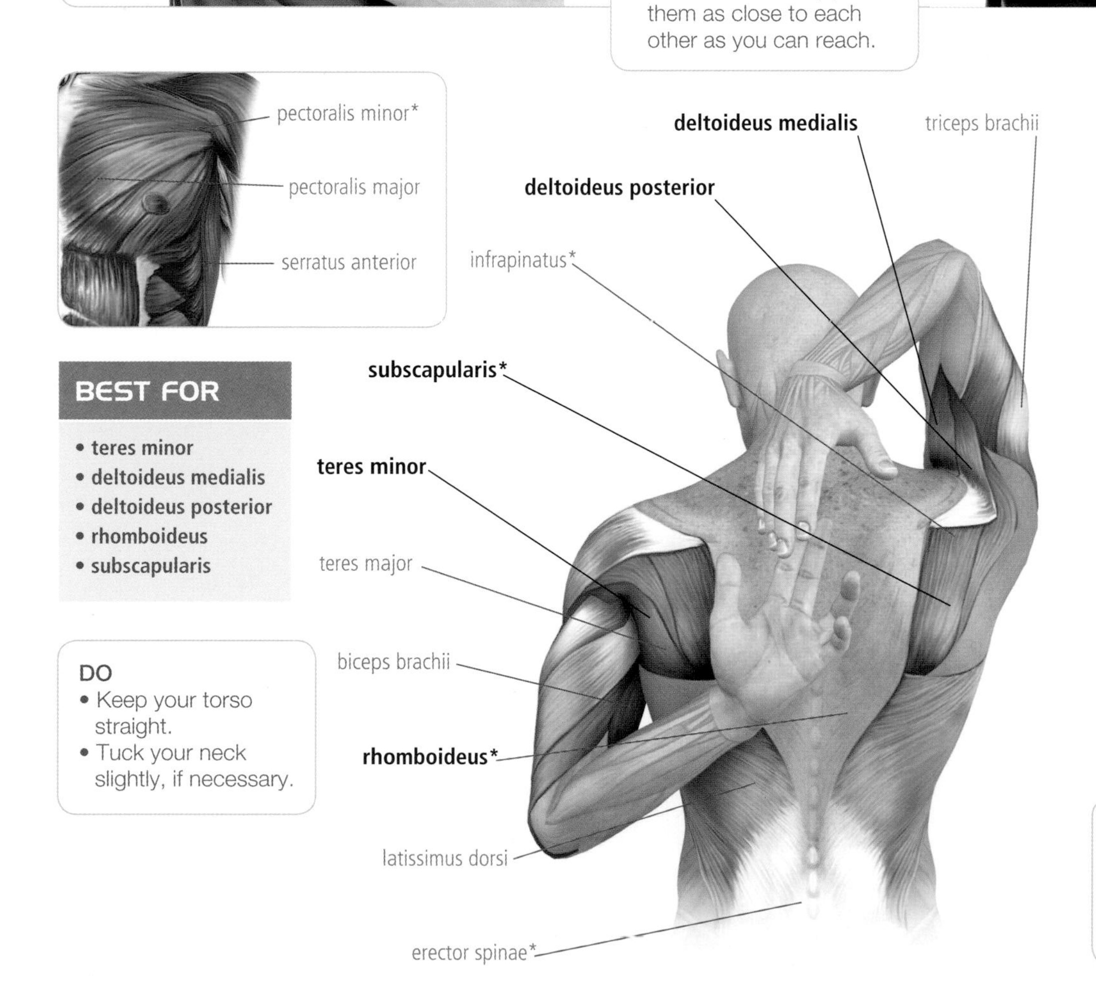

BEST FOR
- **teres minor**
- **deltoideus medialis**
- **deltoideus posterior**
- **rhomboideus**
- **subscapularis**

DO
- Keep your torso straight.
- Tuck your neck slightly, if necessary.

TARGETS
- Shoulders
- Upper back
- Upper arms

LEVEL
- Intermediate

BENEFITS
- Improves shoulders' range of motion

AVOID IF YOU HAVE . . .
- Shoulder injury

ANNOTATION KEY
Bold text indicates target muscles
Gray text indicates other working muscles
* indicates deep muscles

FORWARD ARM REACH

1. Stand upright, gazing forward. Straighten your arms, fingers outstretched, and bring them behind your body.
2. With your palms facing inward, bring both arms forward and raise them as high as you can reach.
3. Lower your arms back to starting position and then repeat, performing 10 repetitions through the full range of motion.

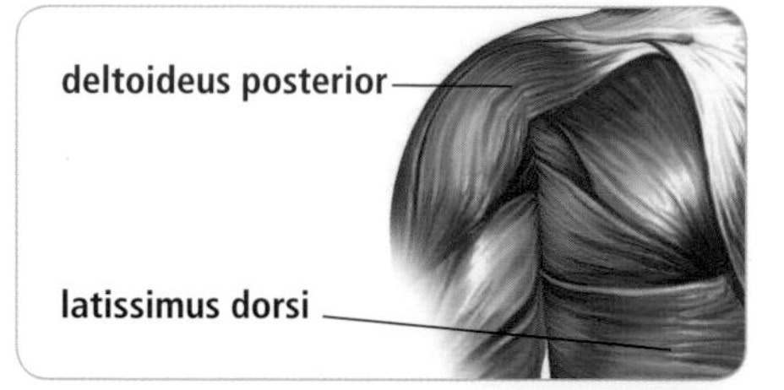

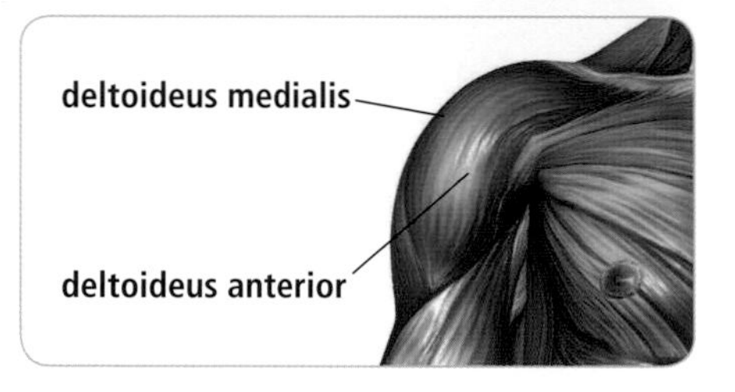

ANNOTATION KEY
Bold text indicates target muscles

TARGETS
- Shoulders

LEVEL
- Beginner

BENEFITS
- Improves shoulders' range of motion

AVOID IF YOU HAVE . . .
- Shoulder injury

BEST FOR

- **deltoideus anterior**
- **deltoideus posterior**
- **deltoideus medialis**
- **latissimus dorsi**

DO
- Keep your arms straight.
- Raise and lower both arms at the same pace.

DON'T
- Move through an excessively wide range of motion.
- Arch you back or bend forward.
- Arch your neck or tuck your chin.

MODIFICATION
Easier: Instead of raising both arms at the same time, use one of your hands to support the opposite arm, lifting only as high as you feel comfortable. Switch arms, and repeat.

MODIFICATION
Harder: Hold a resistance band as you move through your arms' range of motion. Stretch the strap taut to challenge the muscles in your arms.

SIDEWAYS ARM LIFT & CROSS

BEST FOR

- deltoideus anterior
- deltoideus posterior
- deltoideus medialis
- levator scapulae
- rhomboideus
- serratus anterior
- pectoralis major
- pectoralis minor
- intercostales interni
- intercostales externi
- latissimus dorsi

1. Stand upright, gazing forward, with feet planted shoulder-distance apart. Let your arms hang down along your sides.
2. Raise one arm out to the side, and then extend upward as far as you can reach.

DO
- Maintain straight arms throughout the exercise.
- Extend your fingers.
- Keep your torso straight.
- Keep your gaze forward.

3. Lower your arm to the side, so that it is parallel to the floor, and then bring it across your body so that it is pointing in the other side.
4. Return to the starting position, and repeat on the other side.

DON'T
- Lean to one side as you lift the opposite arm.
- Rush through the movement.
- Arch your neck or back.

TARGETS
- Shoulders

LEVEL
- Beginner

BENEFITS
- Improves shoulders' range of motion

AVOID IF YOU HAVE . . .
- Shoulder injury

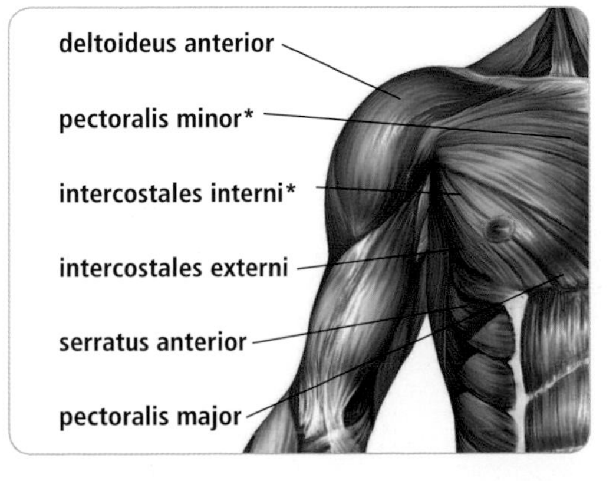

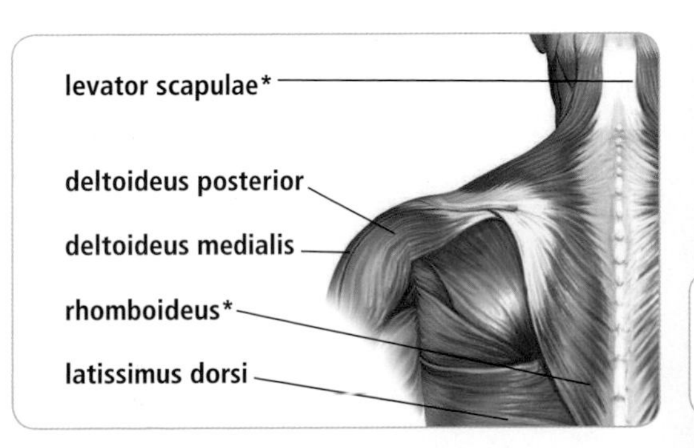

ANNOTATION KEY

Bold text indicates target muscles

* indicates deep muscles

FOREARM EXERCISES

ELBOW BEND & TURN

❶ Begin with your elbows bent at a 90-degree angle, upper arms at your sides and palms facing downward.

❷ Bring your forearms toward your shoulders, turning your palms upward as your move.

❸ Lower your arms to a 90-degree angle. Complete 10 repetitions.

TARGETS
- Elbows

LEVEL
- Beginner

BENEFITS
- Improves elbows' range of motion

AVOID IF YOU HAVE . . .
- Elbow pain

BEST FOR
- **triceps brachii**
- **anconeus**
- **biceps brachii**
- **brachialis**
- **brachioradialis**

ANNOTATION KEY
Bold text indicates target muscles

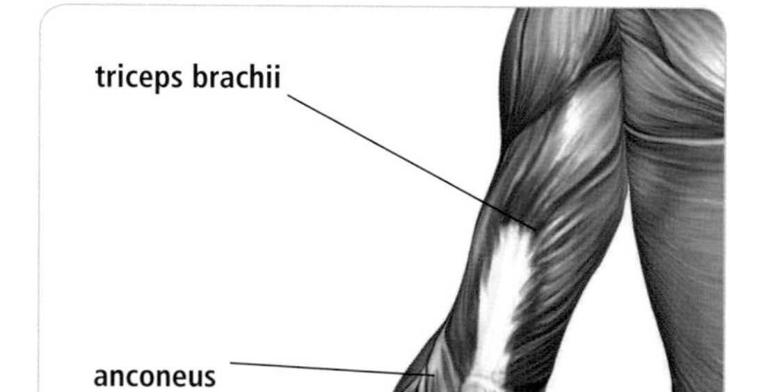

DON'T
- Twist your torso.
- Arch your back or neck.

DO
- Keep your torso straight.
- Gaze forward throughout the exercise.
- Move both arms at the same time and to the same height.
- Keep the rest of your body still as you raise and lower your arms.

biceps brachii

brachialis

brachioradialis

WRIST BEND

1. Sit or stand upright, with your gaze forward. Extend one arm, and position your hand so that the palm is facing forward, with fingers pointing upward.
2. Reverse the motion so that your palm is facing your body, fingers pointing downward.
3. Complete 10 repetitions, and then switch arms and repeat.

BEST FOR

- extensor carpi radialis
- extensor carpi ulnaris
- extensor digiti minimi
- extensor digitorum
- extensor indicis
- extensor pollicis
- flexor carpi radialis
- flexor carpi ulnaris
- flexor digiti minimi
- flexor digitorum
- palmaris longus
- flexor pollicis

DON'T
- Bend your wrist farther than you feel comfortable going, in either direction.

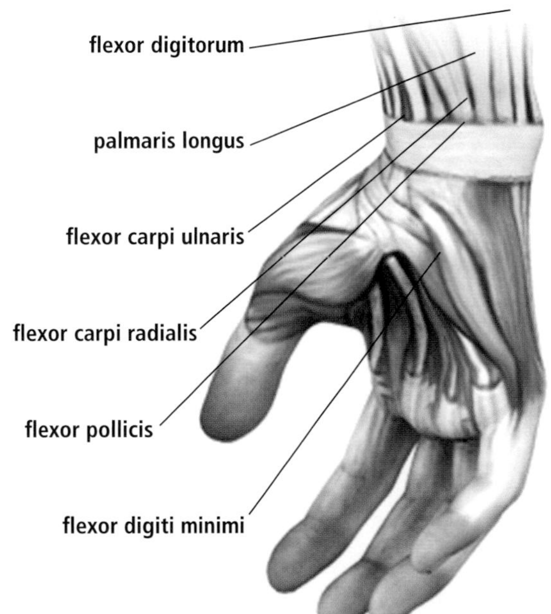

TARGETS
- Wrists

LEVEL
- Beginner

BENEFITS
- Improves wrists' range of motion

AVOID IF YOU HAVE . . .
- Wrist pain

MODIFICATION
Easier: Use your opposite hand to help you stretch in both directions.

DO
- Keep your arms straight.
- Bend at the wrist.

LEG EXERCISES

KNEE LIFT

❶ Sit upright on a chair, with your feet planted on the floor, your hands on your knees, and your gaze forward.

❷ Lift one foot a few inches off the floor. Hold for 5 seconds.

❸ Lower your foot, returning the knee to its 90-degree–angle starting position. Repeat on the other side, and continue to alternate, performing two sets of 10 per side.

DON'T
- Lift your knee to a height that feels uncomfortable.
- Twist your torso.
- Arch your back.

DO
- Keep your body torso straight.
- Keep the rest of your body still as you move one leg at a time.

BEST FOR
- **biceps femoris**
- **semitendinosus**
- **semimembranosus**
- **rectus femoris**
- **vastus lateralis**
- **vastus intermedius***

TARGETS
- Knees

LEVEL
- Beginner

BENEFITS
- Improves knees' range of motion

AVOID IF YOU HAVE . . .
- Knee pain

ANNOTATION KEY
Bold text indicates target muscles
indicates deep muscles*

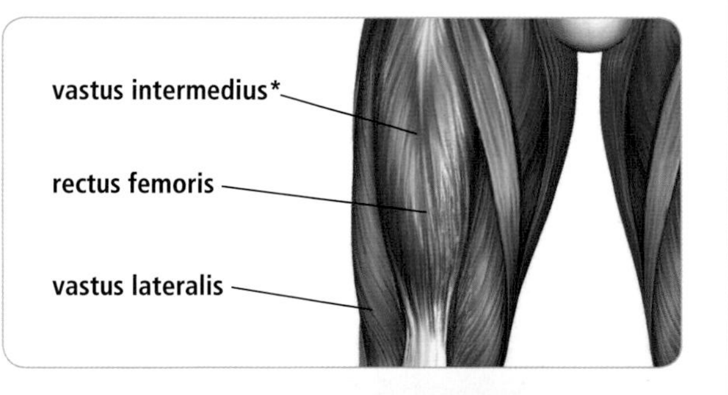

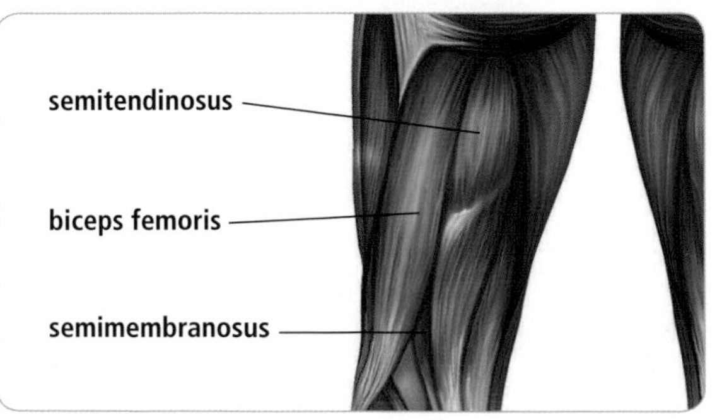

IN-OUT ROTATION

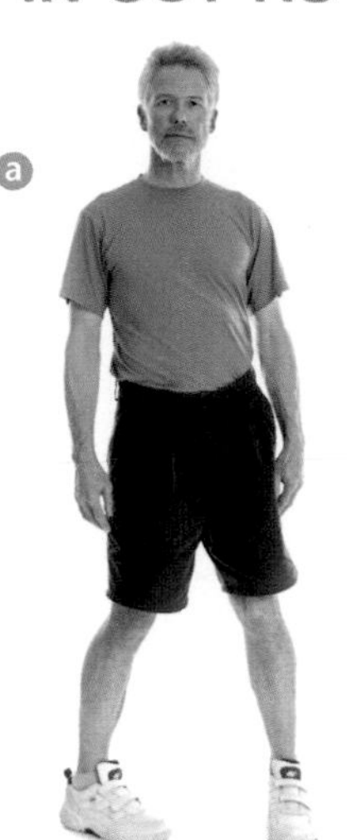

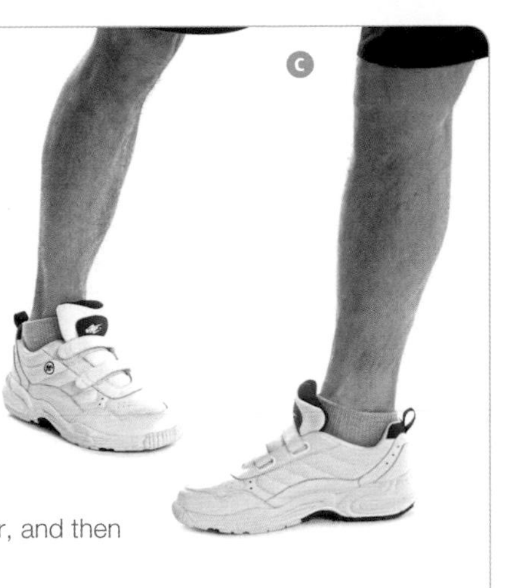

1. Stand with your feet planted a little more than hip-width apart. Place the majority of your weight on one leg.
2. Turn the foot of your other leg inward, rotating from the hip.
3. Return your foot to center, and then rotate it outward.
4. Return to center and repeat on the other side. Continue to alternate feet, performing two sets of 10 per side.

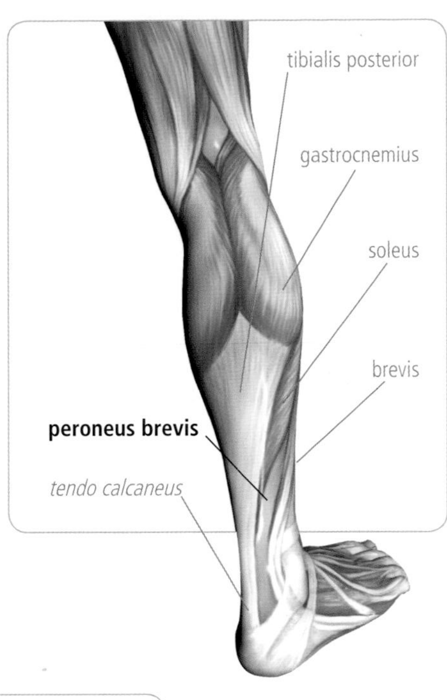

DON'T
- Lean to one side.
- Turn your foot too far inward or outward; go only as far as you feel comfortable.

BEST FOR
- **peroneus brevis**

DO
- Keep your torso straight.
- Make sure that the rotation comes from your hip.
- Keep the rest of your body still as your hip and leg move.

IN-OUT ROTATION TARGETS
- Hips

LEVEL
- Beginner

BENEFITS
- Improves hips' range of motion

AVOID IF YOU HAVE . . .
- Knee pain

ANNOTATION KEY
Bold text indicates target muscles
Gray text indicates other working muscles
Italic text indicates tendons

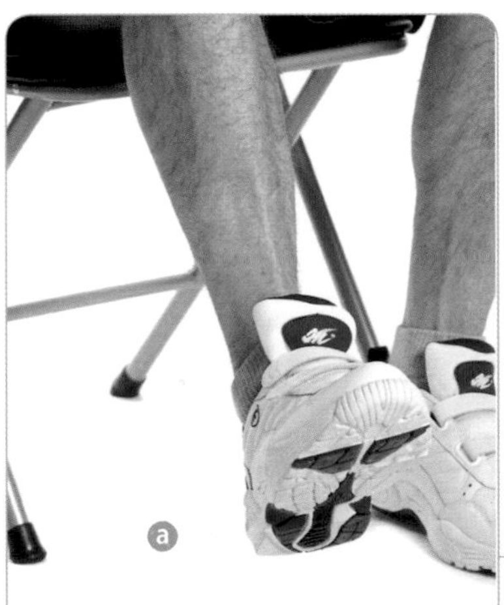

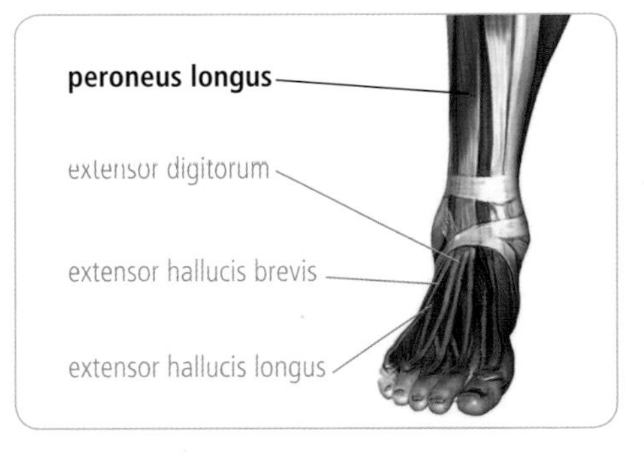

BEST FOR
- **peroneus longus**

ANKLE ROTATION

1. Sit upright with your feet planted close together.
2. Raise one foot a few inches off the floor and rotate it in inward. Pause at the top of the circle, and then rotate your foot outward.
3. Lower your foot and repeat on the other side. Continue to alternate, performing two sets of 10 per side.

DO
- Rotate your foot from your ankle.
- Sit up straight.
- Keep the rest of your body still as your ankle and foot rotate.

DON'T
- Twist your torso.
- Lean over to one side.

ANKLE ROTATION TARGETS
- Ankles

LEVEL
- Beginner

BENEFITS
- Improves ankles' range of motion

AVOID IF YOU HAVE . . .
- Knee pain

CORE-STRENGTHENING EXERCISES

The core is the powerhouse of the body. Major core muscles include the abs, the obliques, and the muscles that support the spine. The core is (quite literally) central to how you look and feel. All bodily movement, in every conceivable direction, originates in the core, and to strengthen it is to guard against injury, improve the body's functionality, and build fitness from the inside out. Effective core strengthening comes in various forms; from the following exercises, choose a regime that works for you.

HIGH PLANK

1. Kneel on all fours, facing downward. Your hands should be planted on the floor, shoulder-width apart, and your knees bent at right angles.

DO
- Keep your arms extended throughout the exercise.
- Engage your abs, keeping them pulled in as you hold the position.
- Keep your neck straight and your gaze downward.
- Start by holding for just 15 seconds, if desired.

BEST FOR
- erector spinae
- transversus abdominis
- rectus abdominis
- obliquus externus
- obliquus internus

TARGETS
- Abs
- Back
- Obliques

LEVEL
- Beginner

BENEFITS
- Strengthens and stabilizes core

AVOID IF YOU HAVE . . .
- Shoulder issues
- Wrist injury
- Lower-back injury

2. Straighten your legs and come onto your toes so that your body forms a line. Hold for 30 seconds, building up to 2 minutes if desired.

DON'T
- Arch your neck by trying to look forward.
- Arch your back or allow it to curve forward.

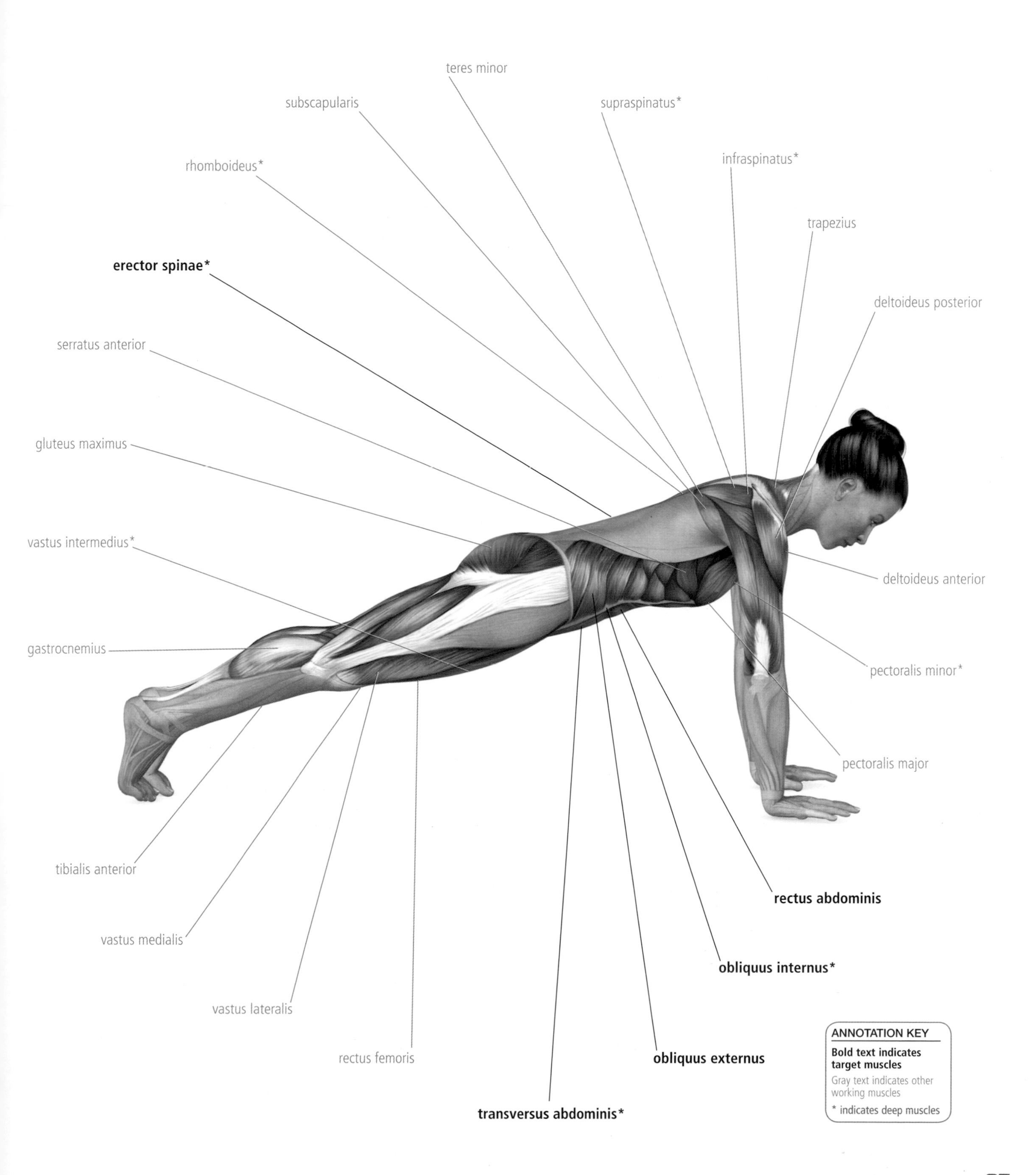
teres minor
subscapularis
supraspinatus*
infraspinatus*
rhomboideus*
trapezius
erector spinae*
deltoideus posterior
serratus anterior
gluteus maximus
vastus intermedius*
deltoideus anterior
gastrocnemius
pectoralis minor*
pectoralis major
tibialis anterior
rectus abdominis
vastus medialis
obliquus internus*
vastus lateralis
rectus femoris
obliquus externus
transversus abdominis*
ANNOTATION KEY
Bold text indicates target muscles
Gray text indicates other working muscles
* indicates deep muscles

PLANK

(a)

1. Lie supine on the floor, with your legs extended behind you. Bend your arms so that your forearms and palms rest flat on the floor.

DON'T
- Allow your shoulders to collapse into your shoulder joints.
- Arch your neck.
- Allow your back to sag.

(b)

2. Bend your knees, supporting your weight between your knees and your forearms, and then push through with your forearms to bring your shoulders up toward the ceiling as you straighten your legs.

3. With control, lower your shoulders until your feel them coming together at your back. Hold for 30 seconds, building up to 2 minutes if desired.

TARGETS
- Abs
- Back
- Obliques

LEVEL
- Beginner

BENEFITS
- Strengthens and stabilizes core

AVOID IF YOU HAVE . . .
- Shoulder injury
- Severe back pain

BEST FOR
- **erector spinae**
- **transversus abdominis**
- **rectus abdominis**
- **obliquus externus**
- **obliquus internus**

ANNOTATION KEY

Bold text indicates target muscles

Gray text indicates other working muscles

* indicates deep muscles

DO
- Keep your abs tight.
- Keep your body in a straight line.
- Lengthen through your neck.
- Start by holding for just 15 seconds, if desired.

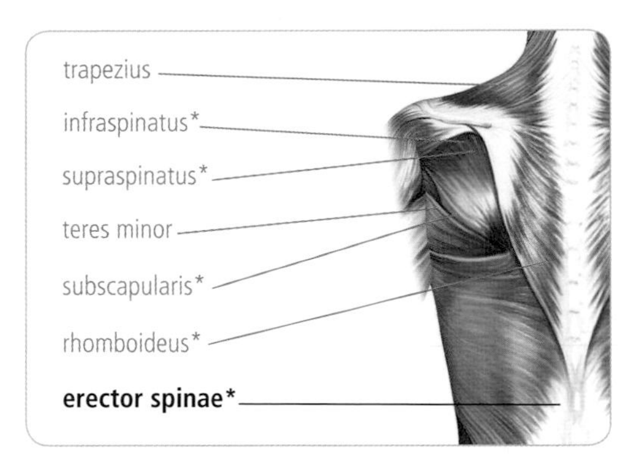

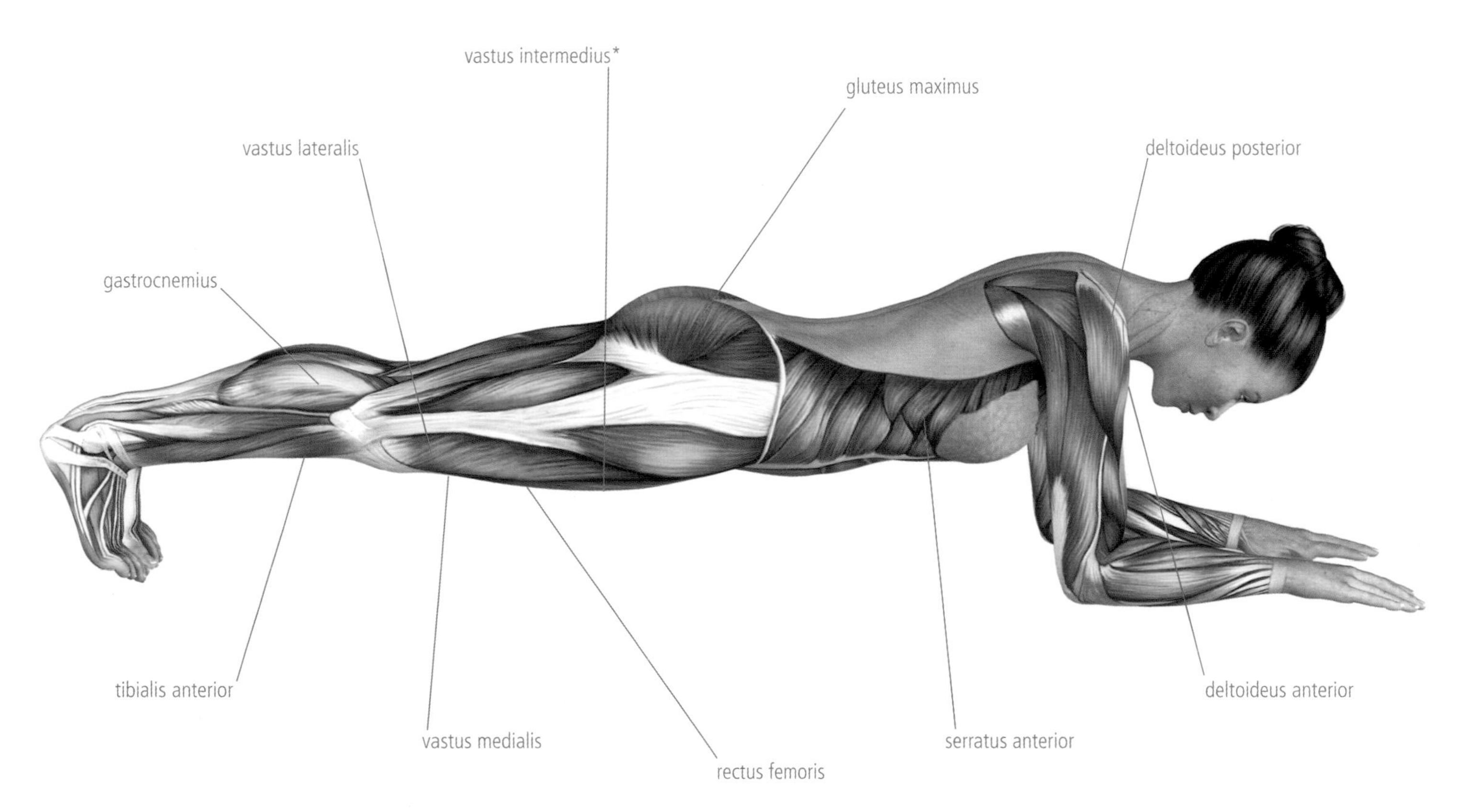

MODIFICATION

Harder: While in the plank position, lift and lower your legs one at a time. Keep the rest of your body still and your abs engaged throughout.

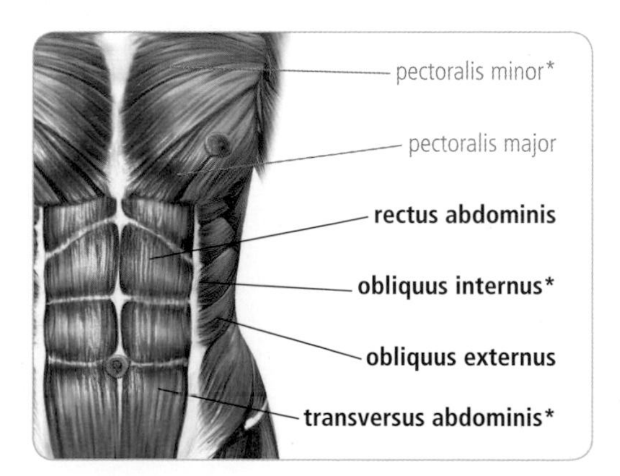

SWISS BALL PLANK WITH LEG LIFT

1. Kneel on your hands and knees, with a Swiss ball behind you. Your hands should be planted on the floor, with your arms straight.

a

DON'T
- Allow your back to sag.
- Arch your neck.
- Lift your legs too high; raise them only as high as you can maintain your form, keeping the rest of your body stable.

2. One at a time, place your feet on the ball so that your legs are fully extended behind you and your body forms a line from head to toe. Find your balance.

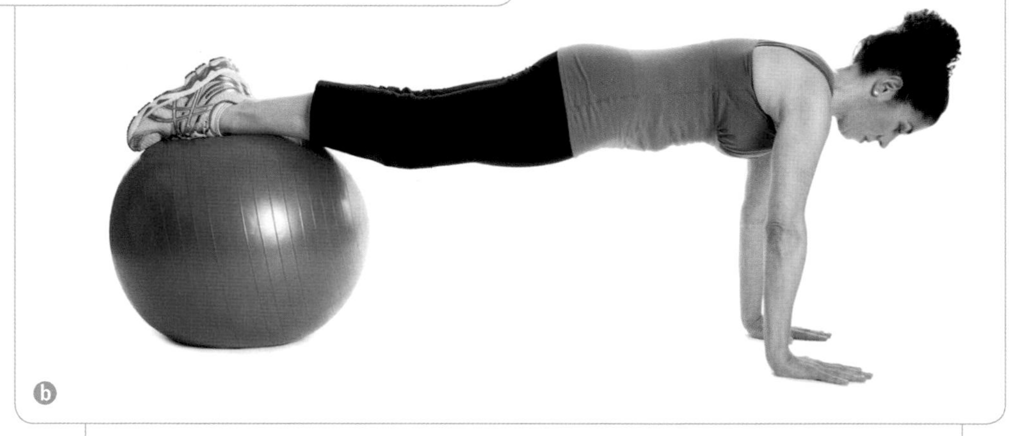

b

TARGETS
- Abs
- Back
- Glutes
- Obliques

LEVEL
- Intermediate

BENEFITS
- Strengthens and stabilizes core

AVOID IF YOU HAVE . . .
- Shoulder injury
- Wrist pain
- Lower-back issues

3. Slowly and with control, raise one foot off the ball. Hold for as long as you can, starting with 10 to 15 seconds and working your way up to 1 minute.

4. Lower your foot to the ball, and repeat on the other side.

c

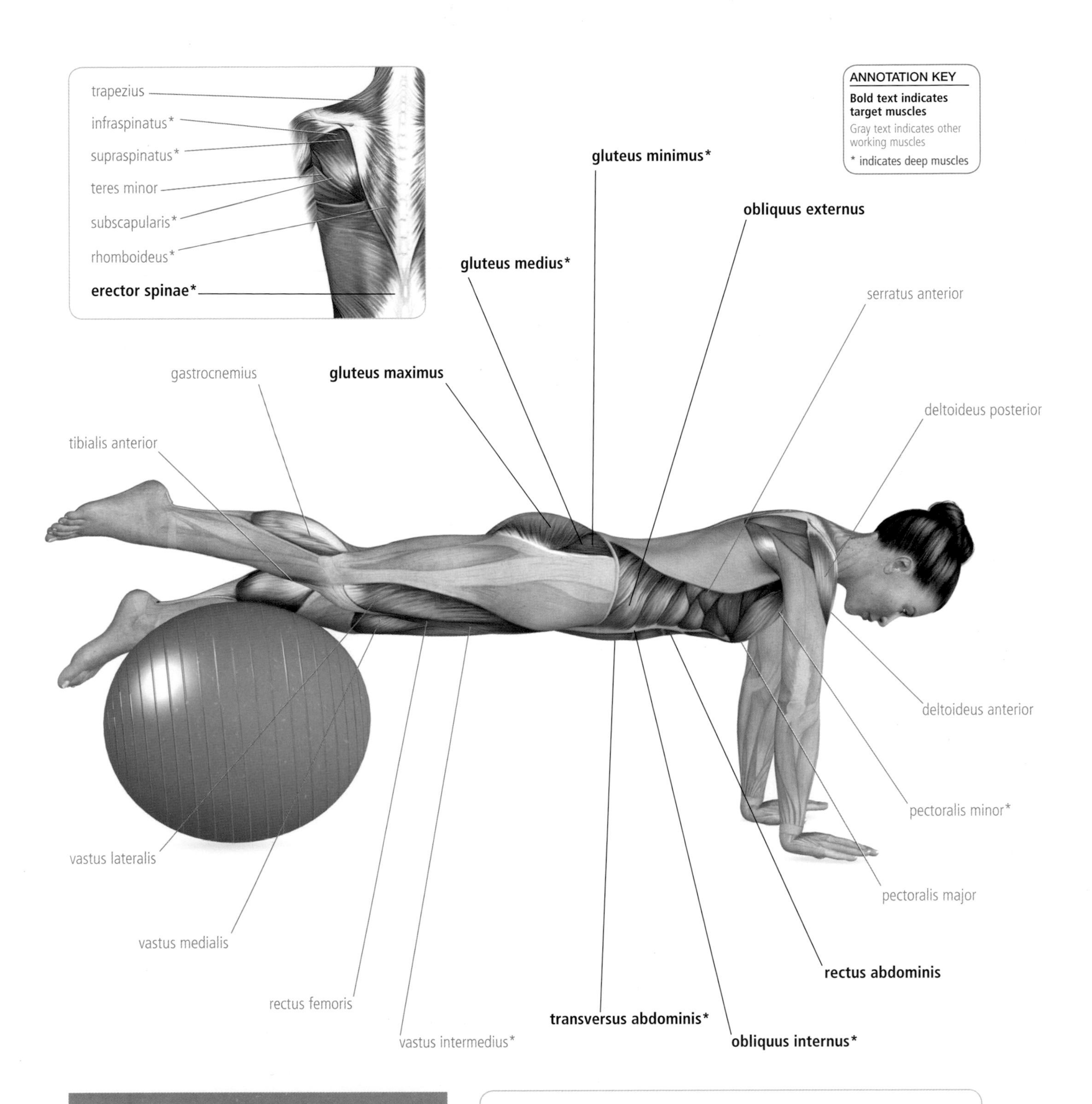

BEST FOR

- erector spinae
- transversus abdominis
- rectus abdominis
- obliquus externus
- obliquus internus
- gluteus maximus
- gluteus minimus
- gluteus medius

DO
- Keep your abs tight.
- Keep your body in a straight line.
- Keep your neck straight and your gaze downward.
- Engage your whole body as you work to keep the ball in place.
- Start by holding for just 15 seconds per leg if desired.

SIDE PLANK

1. Lie on your side, with your legs extended and stacked one on top of the other. Bend one arm to a 90-degree angle with your forearm on the floor, fingers facing forward. Rest your other arm along the side of your body.

DON'T
- Arch your back.
- Allow your hips to sink.
- Arch your neck.
- Hunch your shoulders.

2. Push into the floor with your hand and forearm, and raise your hips off the floor until your body forms a straight line. Hold for 30 seconds, working up to 1 minute.

3. Lower, and repeat on the other side.

TARGETS
- Abs
- Glutes
- Chest
- Obliques
- Shoulders

LEVEL
- Advanced

BENEFITS
- Strengthens abs, lower back, and shoulders
- Stabilizes trunk

AVOID IF YOU HAVE . . .
- Rotator cuff injury
- Neck issues
- Lower-back pain

BEST FOR

- **transversus abdominis**
- **obliquus internus**
- **obliquus externus**
- **rectus abdominis**
- **adductor magnus**
- **adductor longus**
- **gluteus minimus**
- **gluteus medius**
- **pectoralis major**

MODIFICATION

Same level of difficulty:
Cross your legs at the ankles instead of stacking your feet.

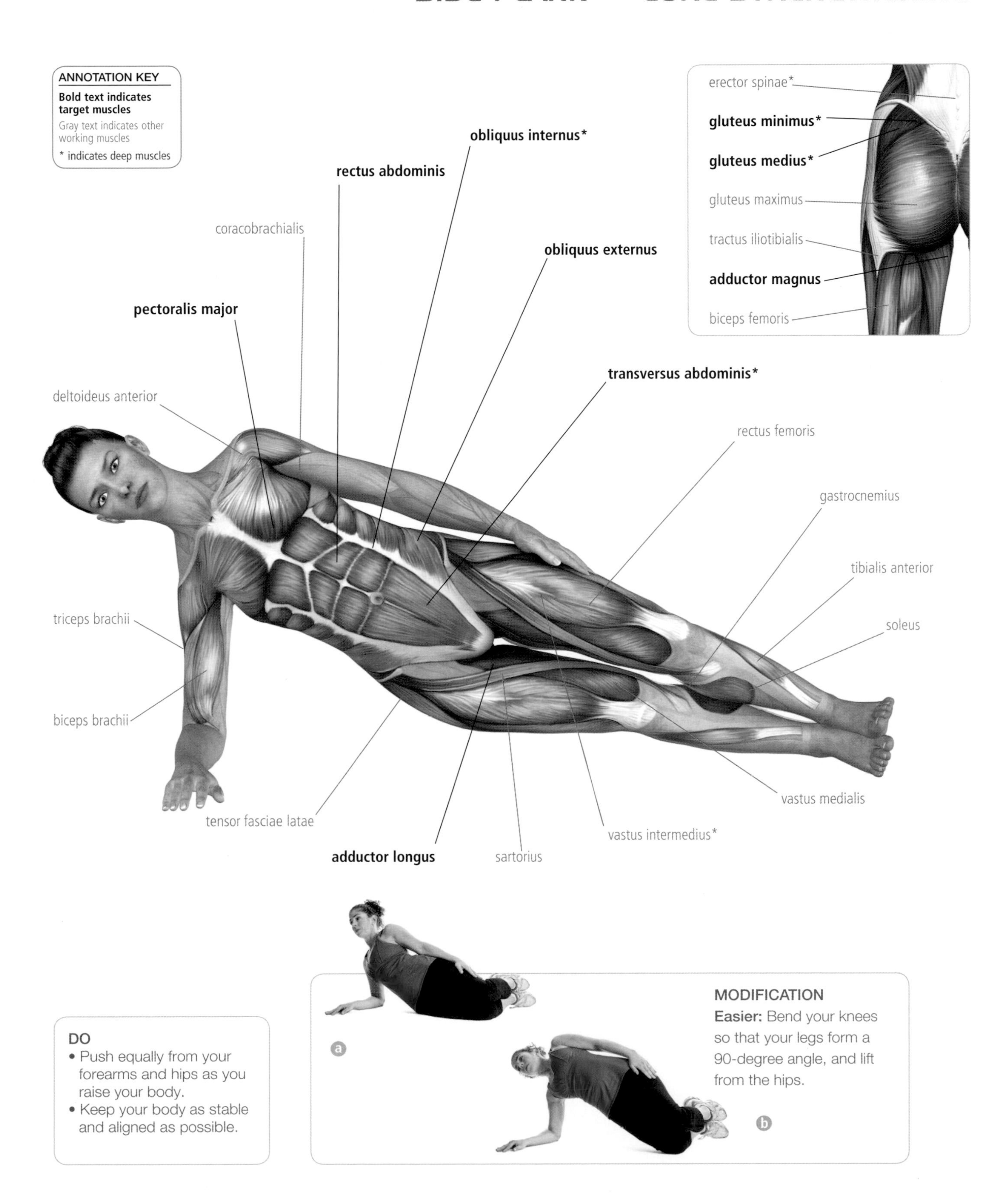

MODIFICATION

Easier: Bend your knees so that your legs form a 90-degree angle, and lift from the hips.

DO

- Push equally from your forearms and hips as you raise your body.
- Keep your body as stable and aligned as possible.

T-STABILIZATION

CORE STRENGTHENING

1 Begin on your side, with your legs stacked one on top of the other, knees slightly bent. Press your hips into the floor, and use both hands to support your torso.

a

BEST FOR

- **rectus abdominis**
- **deltoideus posterior**
- **gluteus maximus**
- **transversus abdominis**

2 With one hand planted directly beneath your shoulder, press your body upward into a side plank.

3 Draw your navel toward your spine, and extend one arm toward the ceiling. Hold for 30 seconds, working up to 1 minute.

4 Lower, switch sides, and repeat.

b

TARGETS
- Abs
- Hips
- Lower back
- Obliques

LEVEL
- Advanced

BENEFITS
- Strengthens abs and shoulders

AVOID IF YOU HAVE . . .
- Lower-back pain
- Shoulder issues
- Wrist pain

DO
- Try to keep your body as stable and aligned as possible.
- Follow your top arm with your gaze.
- Keep your shoulders stable.

DON'T
- Allow your shoulder to sink into its socket.
- Allow your hips to sag.

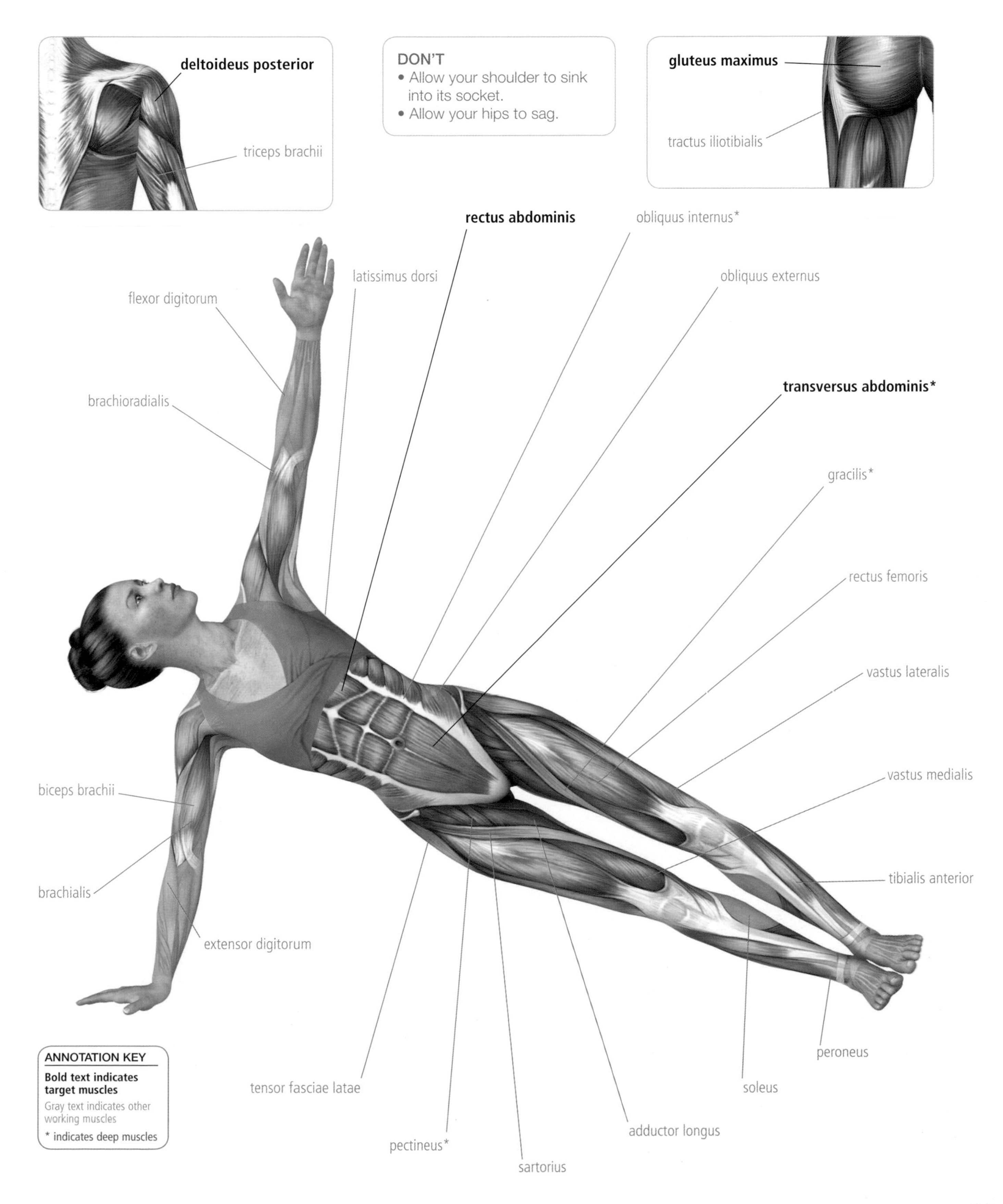

ANNOTATION KEY
Bold text indicates target muscles
Gray text indicates other working muscles
* indicates deep muscles

PRONE COBRA

1. Lie facedown, with your legs extended behind you, feet about shoulder-width apart. Extend your arms along your sides, angled slightly away from your body, with palms downward.

2. Keeping your hands on the floor, raise both your upper and your lower body slightly off the floor.

TARGETS
- Back

LEVEL
- Intermediate

BENEFITS
- Strengthens back muscles

AVOID IF YOU HAVE . . .
- Lower-back issues

3. Raise your arms upward as you lift your upper and lower body farther off the floor.

4. Lower, and repeat for two sets of 15 repetitions.

ANNOTATION KEY
Bold text indicates target muscles
Gray text indicates other working muscles
* indicates deep muscles

DON'T
- Swing your arms upward in a vigorous manner; instead, move smoothly and with control.
- Arch your neck.

DO
- Keep your neck straight, gazing downward.
- Squeeze your buttocks as you lift and lower.
- If desired, hold the raised position for a few seconds before lowering.

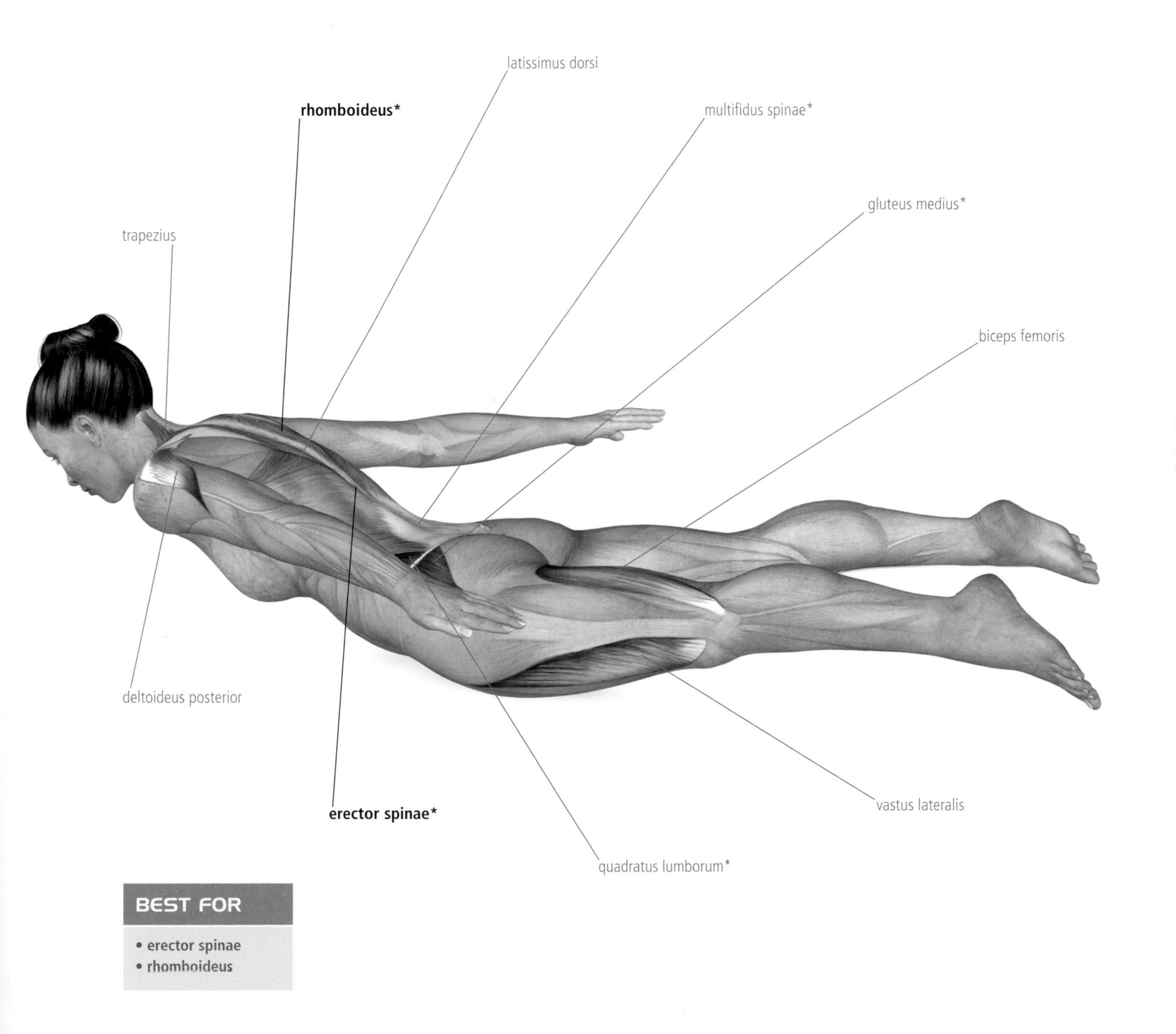

BEST FOR
- erector spinae
- rhomboideus

SWISS BALL JACKKNIFE

1. Kneel on your hands and knees, with a Swiss ball behind you. Your hands should be planted on the floor, with your arms straight.

2. One at a time, place your feet on the ball so that your legs are fully extended behind you and your body forms a line from head to toe. Find your balance.

3. Flex your hips, and pull your knees toward your chest, driving your hips toward the ceiling and retracting your abdomen.

TARGETS
- Hip flexors
- Upper abs

LEVEL
- Advanced

BENEFITS
- Stabilizes core
- Strengthens and tones abs

AVOID IF YOU HAVE . . .
- Lower-back issues
- Neck pain

BEST FOR
- rectus abdominis
- rectus femoris
- iliopsoas
- pectineus
- tensor fasciae latae

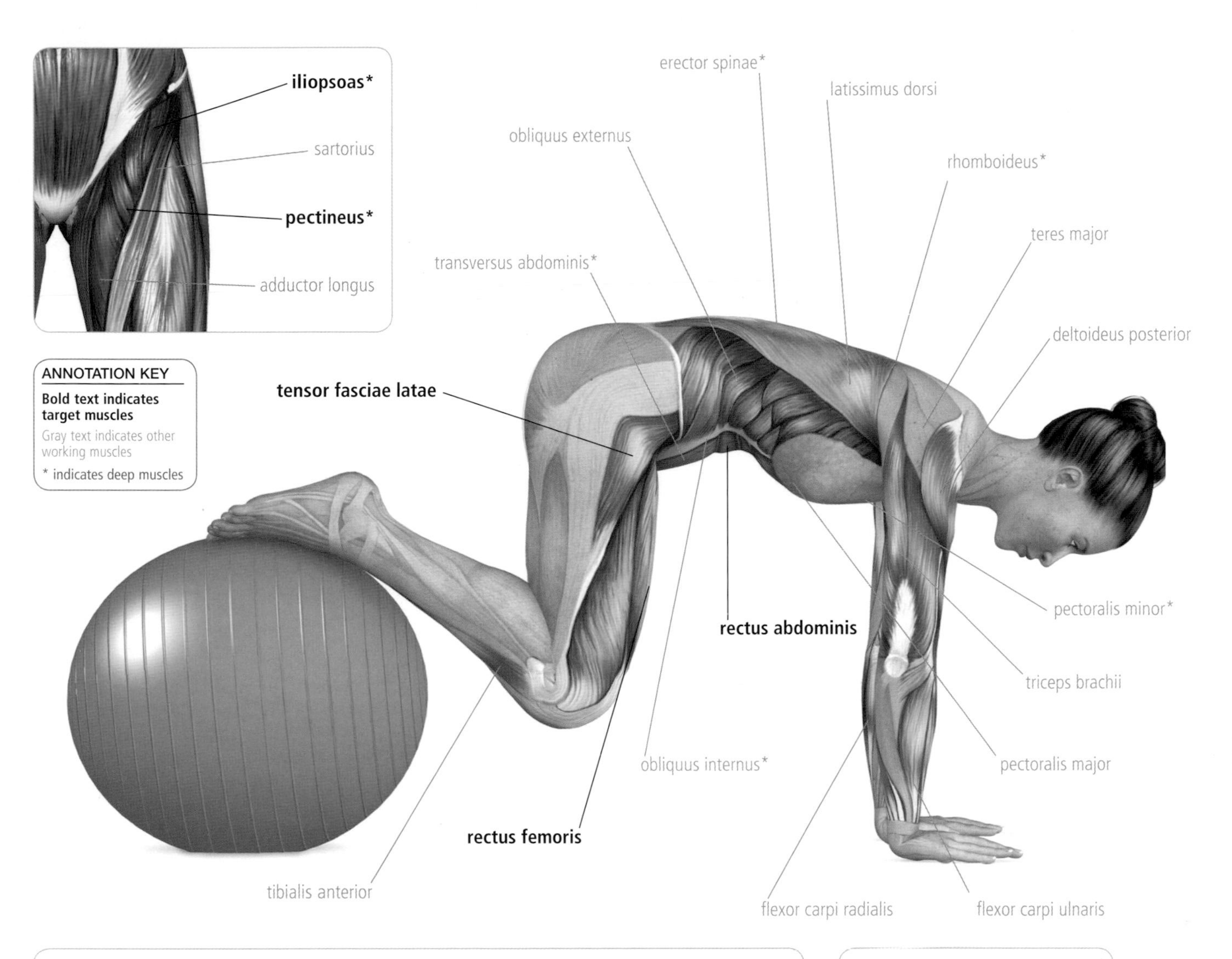

4. Continue to pull inward until your glutes are resting on your heels.

5. Hold for 5 seconds, and then straighten your legs to the starting position. Begin with 10 repetitions, working up to 20.

DO
- Engage your core, keeping your abs pulled in.
- Keep your back as straight as possible.
- Keep your body as stable as possible as you move through the exercise.
- When your legs are extended on the ball, keep your legs, torso, and neck in a straight line.
- Keep your gaze downward.

DON'T
- Arch your back or neck.
- Round your back forward.

SWISS BALL ROLLOUT

BEST FOR
- rectus abdominis
- erector spinae

1 Kneel in front of your Swiss ball, with your hands resting on the ball.

a

DO
- Keep your upper body elongated.
- Keep your lower legs and feet anchored to the floor throughout the exercise.
- Maintain a flat back.
- Keep your abs pulled in.
- Move smoothly and with control.

TARGETS
- Back
- Upper abs

LEVEL
- Intermediate

BENEFITS
- Stabilizes core

AVOID IF YOU HAVE . . .
- Lower-back issues
- Knee issues

2 Use your hands to roll the ball slightly in front of you as you begin to lean forward.

b

DON'T
- Allow your hips to sag.

3. Leading with your arms and following with your body, roll the ball farther forward.

4. Using your abdominals and lower back, roll back to your starting position. Repeat, working up to three sets of 15.

ANNOTATION KEY

Bold text indicates target muscles

Gray text indicates other working muscles

* indicates deep muscles

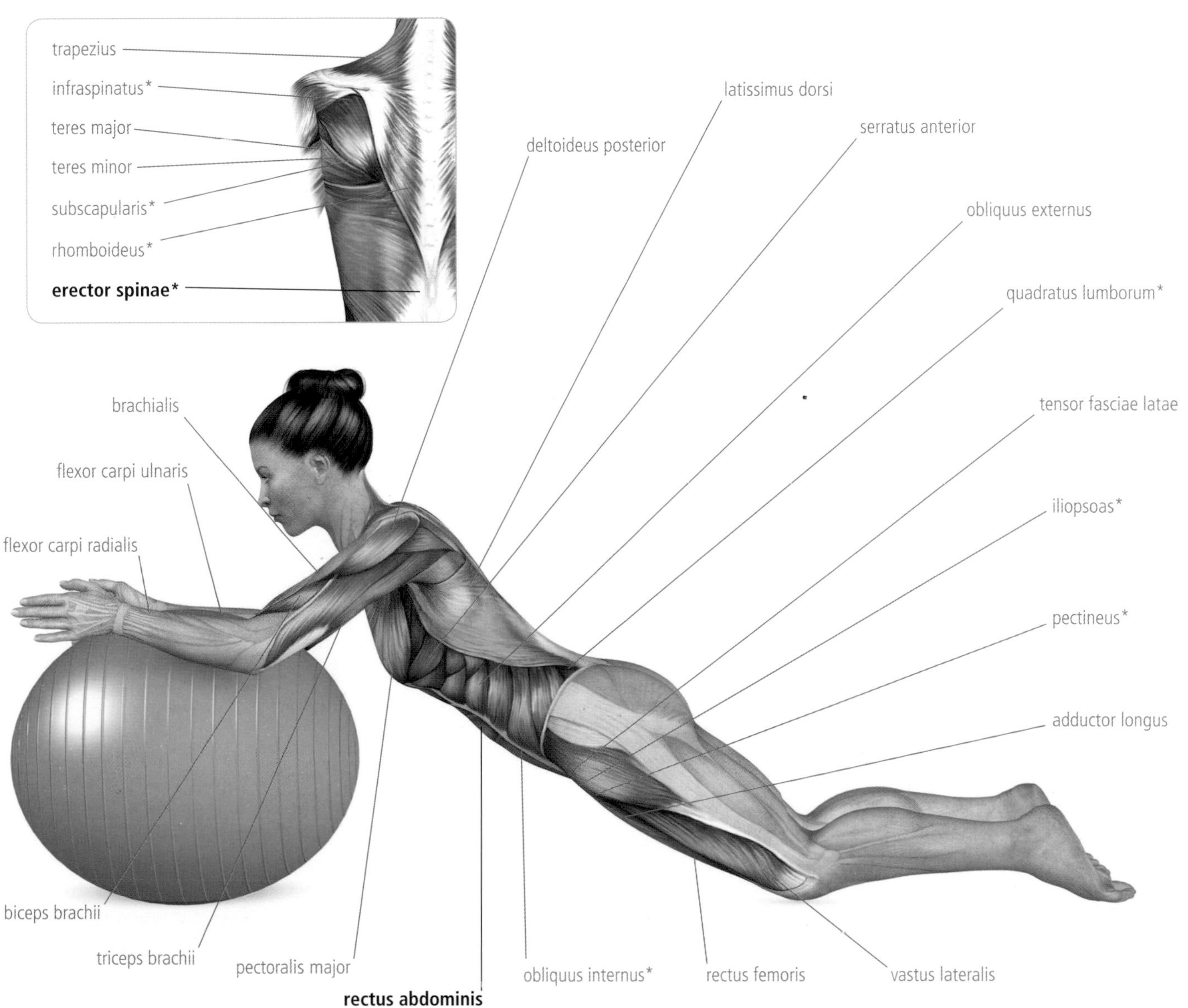

CRUNCH

1. Lie on your back with your legs bent and your hands behind your head, elbows flared outward.
2. Contracting your abdominals, raise your head and shoulders off the floor.

a

3. Lower, and repeat. Complete three sets of 25.

DON'T
- Use your neck to drive the movement.

b

DO
- Lead with your abs, as if a string were hoisting you up by your belly button.
- Try to keep your feet planted on the floor.
- Keep your elbows flared outward.

BEST FOR
- rectus abdominis
- transversus abdominis

TARGETS
- Abs

LEVEL
- Beginner

BENEFITS
- Strengthens and helps to define abs

AVOID IF YOU HAVE . . .
- Lower-back issues
- Neck pain

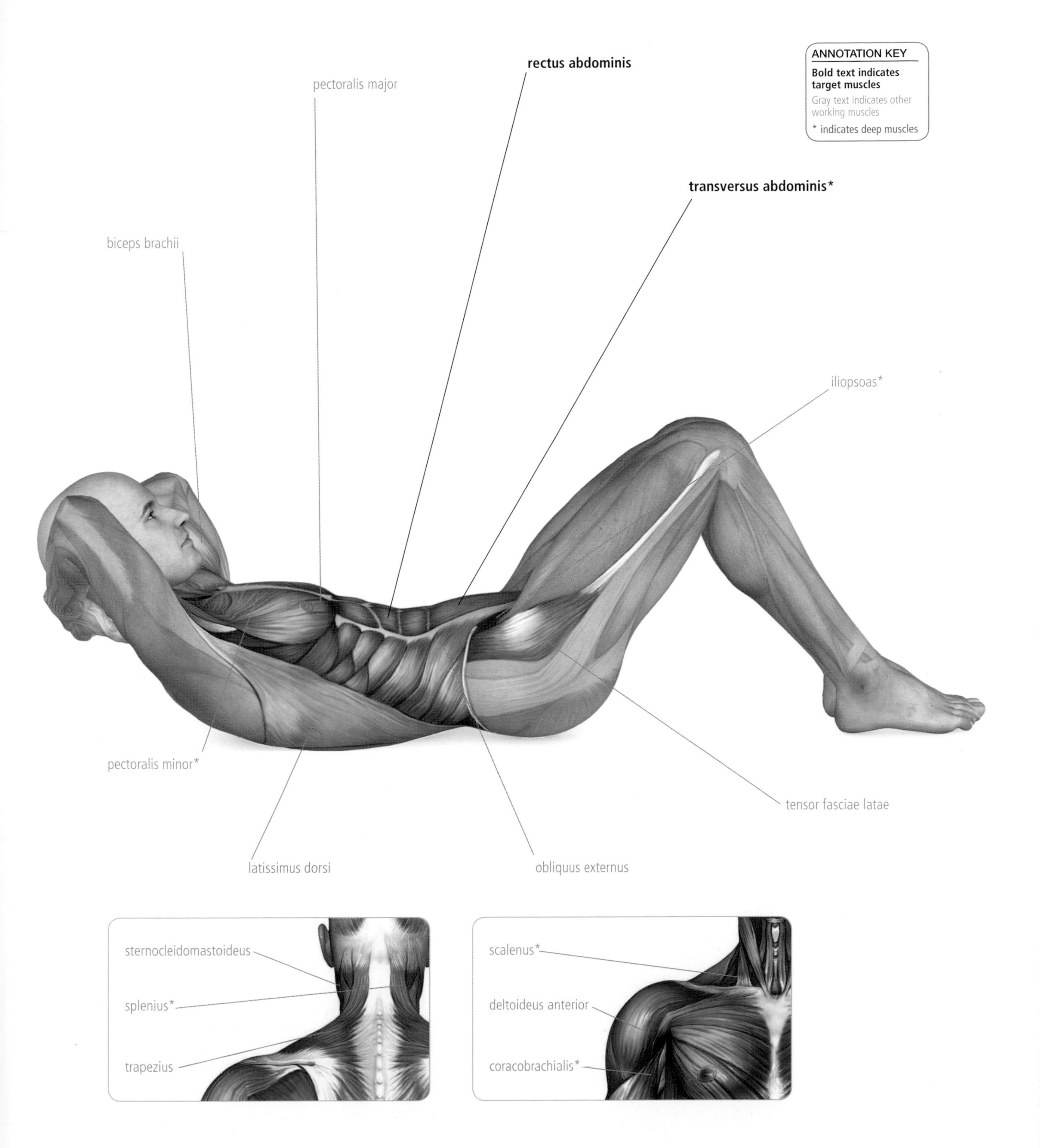
ANNOTATION KEY
Bold text indicates target muscles
Gray text indicates other working muscles
* indicates deep muscles
rectus abdominis
pectoralis major
transversus abdominis*
biceps brachii
iliopsoas*
pectoralis minor*
tensor fasciae latae
latissimus dorsi
obliquus externus
sternocleidomastoideus
splenius*
trapezius
scalenus*
deltoideus anterior
coracobrachialis*

REVERSE CRUNCH

1. Lie on your back with your arms extended along your sides and your feet off the floor. Your legs should be slightly bent.

a

b

2. Tuck your legs in toward your body as you lift your glutes, and then lower back a few inches off the floor.

3. Lower in a controlled manner, returning your feet to their original position. Repeat, performing three sets of 20.

TARGETS
- Upper abs

LEVEL
- Intermediate

BENEFITS
- Strengthens and helps to define abs

AVOID IF YOU HAVE . . .
- Hip instability
- Lower-back issues

BEST FOR
- **rectus abdominis**
- **transversus abdominis**

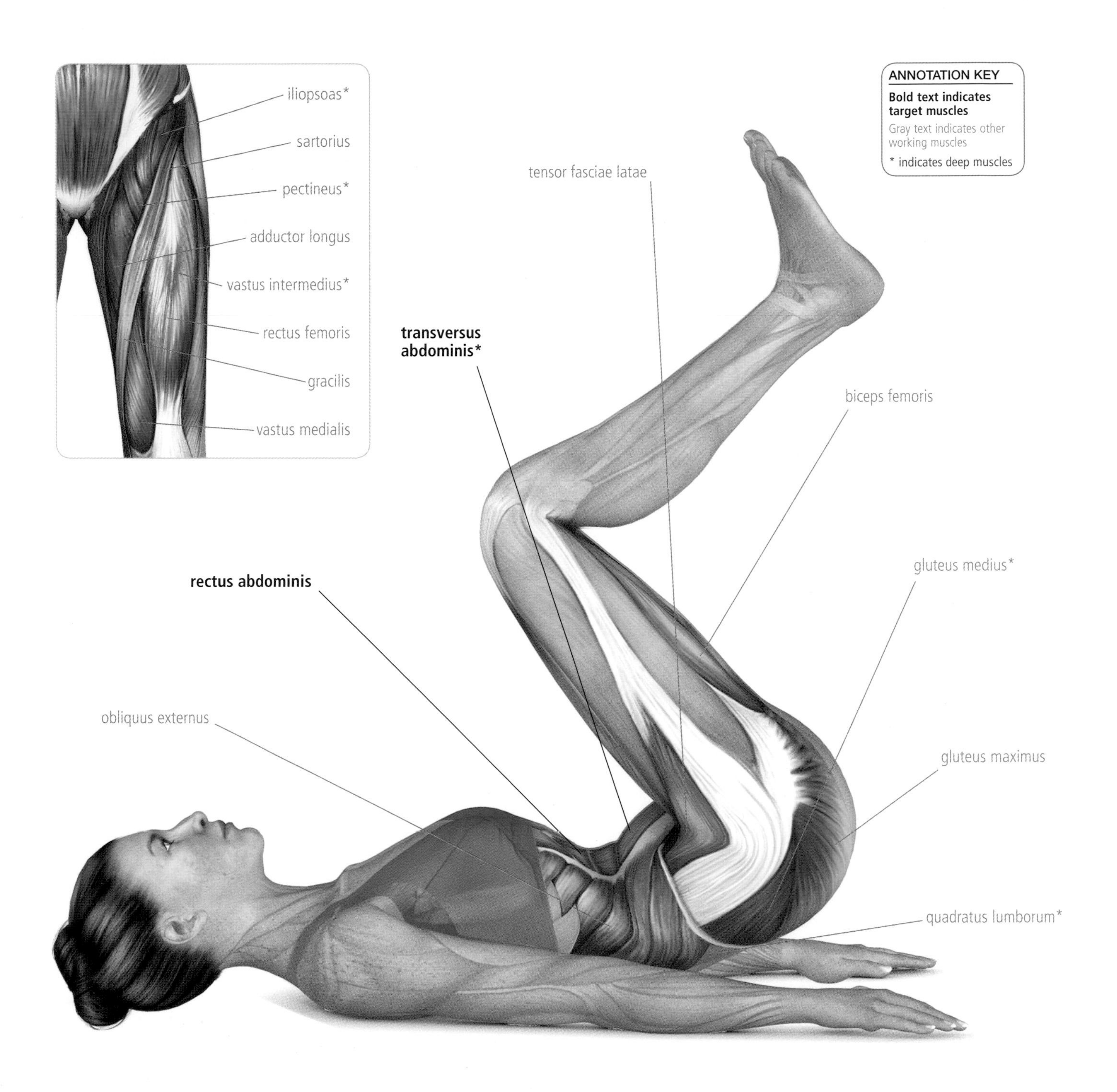

DON'T
- Lift with your lower back or neck.
- Rely on momentum to help you perform the movement.

DO
- Use your abdominals to drive your lower body's movement.
- Keep your arms flat on the floor.

LEG RAISE

a

1. Lie on your back with your arms along your sides. Extend your legs and lift them off the floor, angled away from your body.

DON'T
- Rely on momentum as you lift and lower your legs.
- Use your lower back to drive the movement.

TARGETS
- Lower abs

LEVEL
- Intermediate

BENEFITS
- Strengthens and tones abs

AVOID IF YOU HAVE . . .
- Lower-back issues

b

2. Raise your legs until they are perpendicular to the floor.

3. Lower your legs so that your feet are just above the floor, and then raise them again, performing two sets of 20.

DO
- Keep your upper body braced.
- Use your abs to drive the movement.
- Move your legs together, as if they were a single leg.
- Keep your arms on the floor.
- Bend your legs—which lessens stress on abs.

BEST FOR

- **rectus abdominis**
- **transversus abdominis**

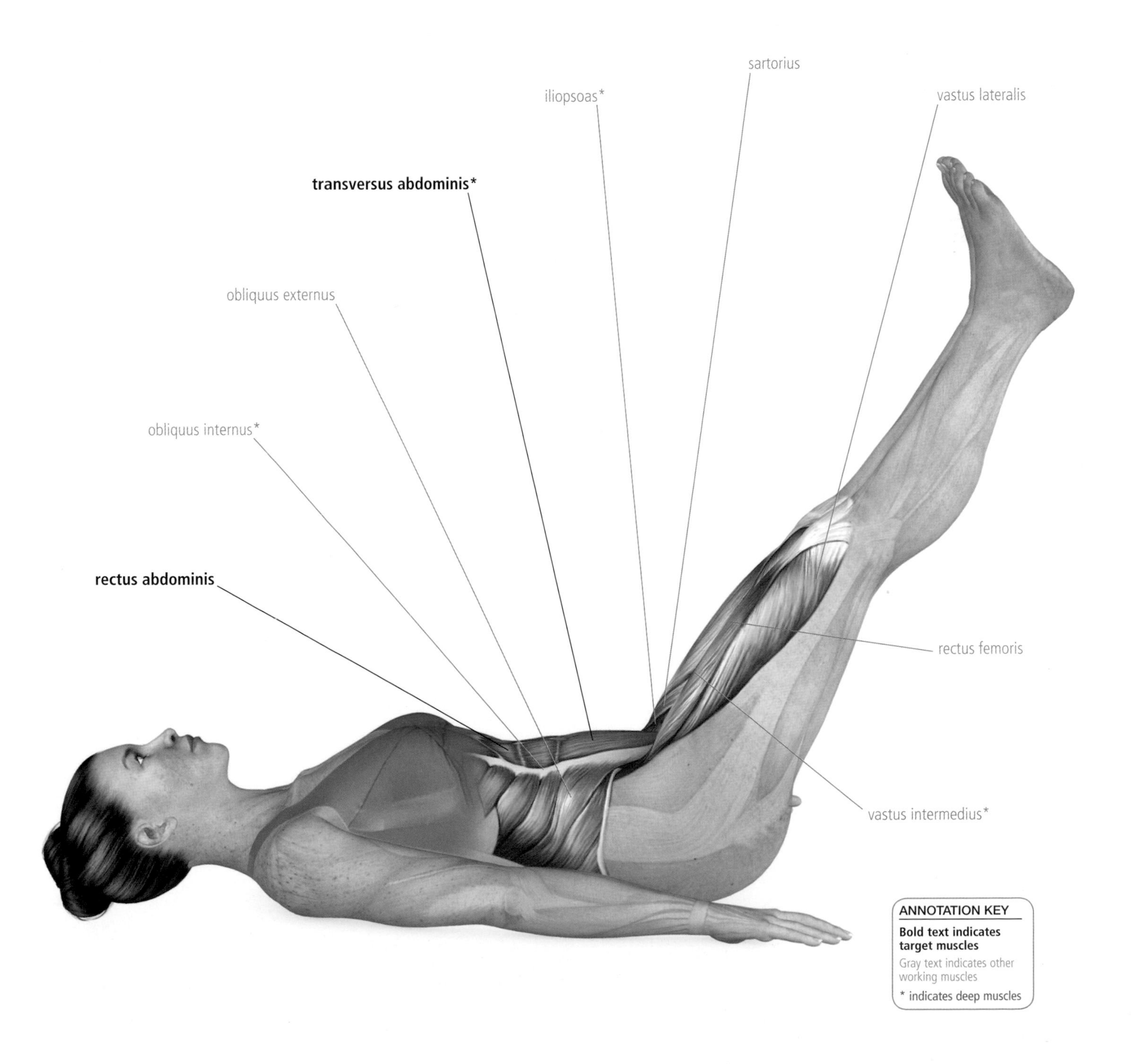

ANNOTATION KEY
Bold text indicates target muscles
Gray text indicates other working muscles
* indicates deep muscles

SEATED RUSSIAN TWIST

CORE STRENGTHENING

a

❶ Sit upright with your legs bent, feet flat on the floor. Extend your arms straight ahead, and lean back slightly to activate your core.

DON'T
- Rush through the twist.
- Shift your feet or knees to the side as you twist.

b

❷ In a smooth motion, rotate your upper body to the side, and then return to center. Repeat rotation on the other side.

❸ Return to center, and repeat the full twist, performing three sets of 20.

TARGETS
- Back
- Obliques
- Upper abs

LEVEL
- Intermediate

BENEFITS
- Stabilizes and strengthens core

AVOID IF YOU HAVE . . .
- Lower-back issues

BEST FOR
- **rectus abdominis**
- **obliquus externus**
- **obliquus internus**
- **erector spinae**
- **transversus abdominis**

DO
- Twist smoothly and with control.
- Keep your back flat as you twist.
- Keep your feet on the floor.
- Keep your arms straight.

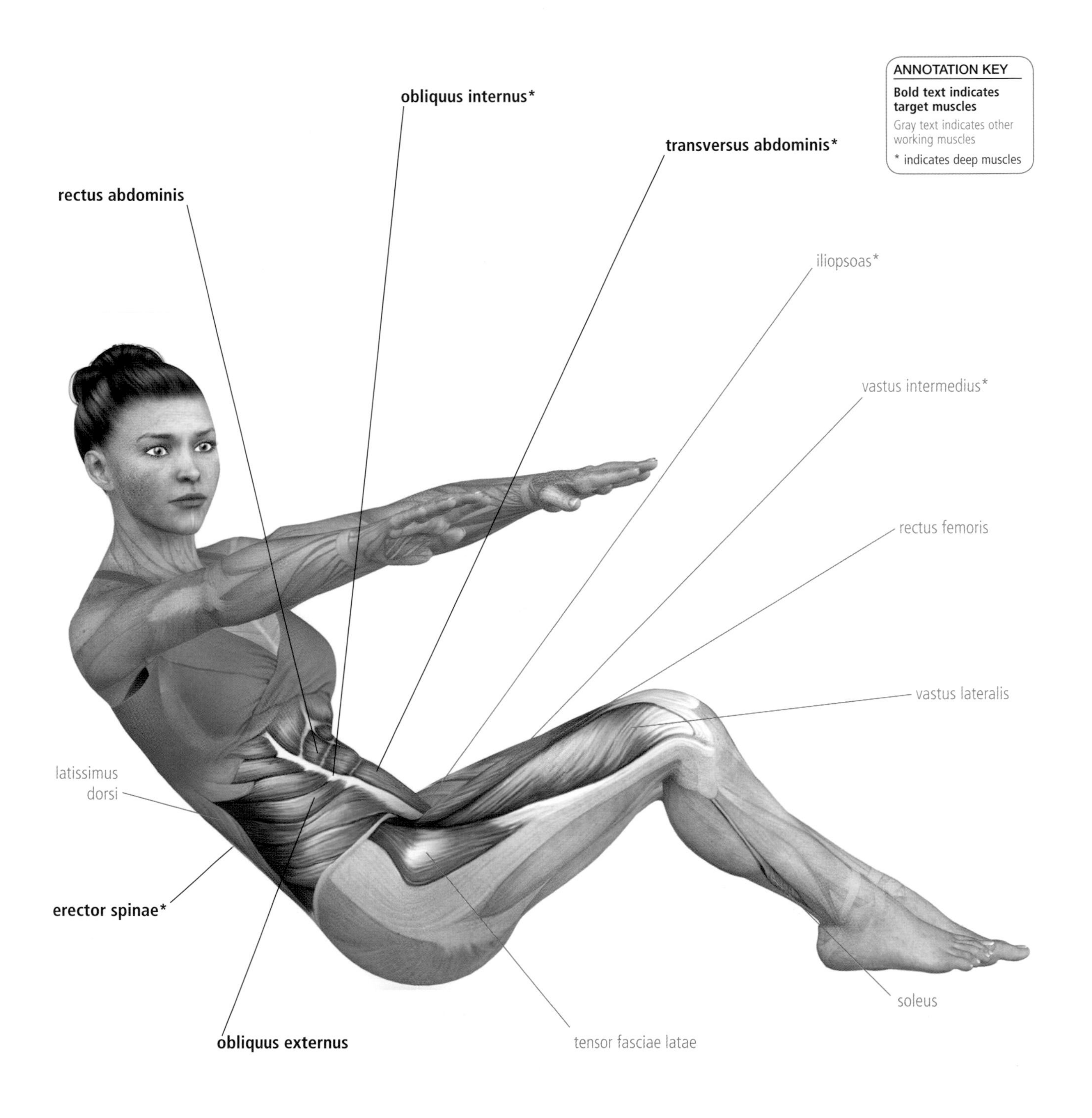

MODIFICATION

Harder: Perform twists holding a medicine ball.

BICYCLE CRUNCH

1. Lie on your back with your fingers at your ears, your elbows flared outward, and your legs bent at a 90-degree angle.

DO
- Use your core to drive the movement.
- Keep your elbows flared.
- Keep both hips stable on the floor.
- Keep your neck elongated.

BEST FOR
- **rectus abdominis**
- **obliquus internus**
- **obliquus externus**

2. Roll up with your torso, reaching one elbow diagonally toward the opposite knee. At the same time, extend the other leg forward.

TARGETS
- Obliques
- Upper abs

LEVEL
- Intermediate

BENEFITS
- Stabilizes core
- Strengthens and tones obliques and upper abdominals

AVOID IF YOU HAVE . . .
- Lower-back issues
- Neck issues

3. Release, and repeat on the other side. Continue to alternate, completing 30 crunches in both directions.

MODIFICATION
Same level of difficulty: Keep both feet flexed throughout the exercise.

DON'T
- Arch your back or raise your lower back off the floor.
- Pull your head upward with your hands.

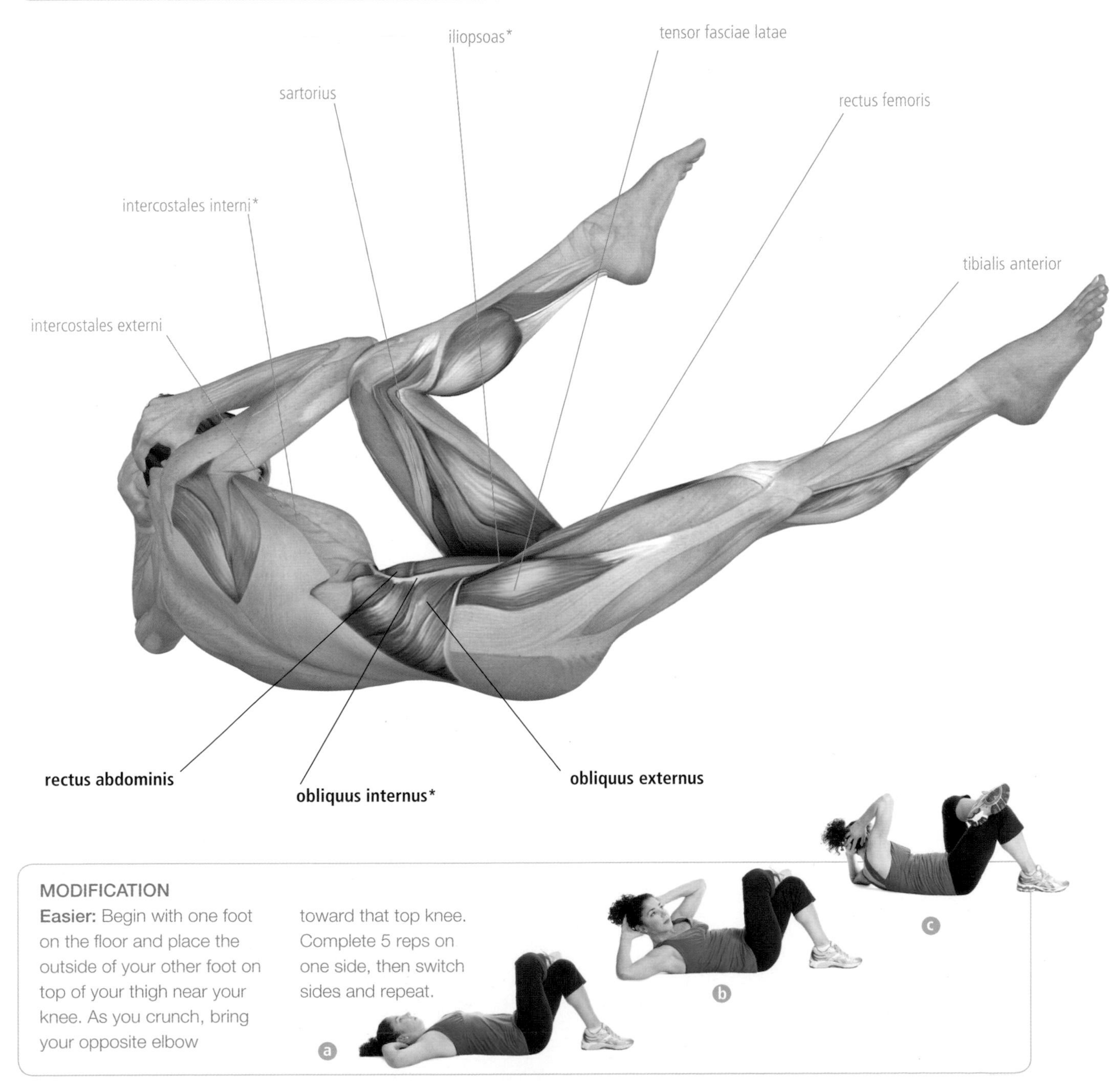

MODIFICATION
Easier: Begin with one foot on the floor and place the outside of your other foot on top of your thigh near your knee. As you crunch, bring your opposite elbow toward that top knee. Complete 5 reps on one side, then switch sides and repeat.

a

b

c

SWISS BALL HIP CROSSOVER

1. Lie on your back, with your arms extended out to your sides. Place your legs on a Swiss ball, with glutes close to the ball.

BEST FOR

- erector spinae
- obliquus externus

2. Brace your abs, and lower your legs to one side, as close to the floor as you can possibly go without raising your shoulders off the floor.

TARGETS
- Lower back
- Obliques

LEVEL
- Intermediate

BENEFITS
- Helps to strengthen and tone abs
- Improves core stabilization

AVOID IF YOU HAVE . . .
- Lower-back issues

MODIFICATION
Easier: Begin with your legs lifted and bent at a 90-degree angle. Try to keep your upper body as stable as possible as you perform the crossover without the ball, alternating sides.

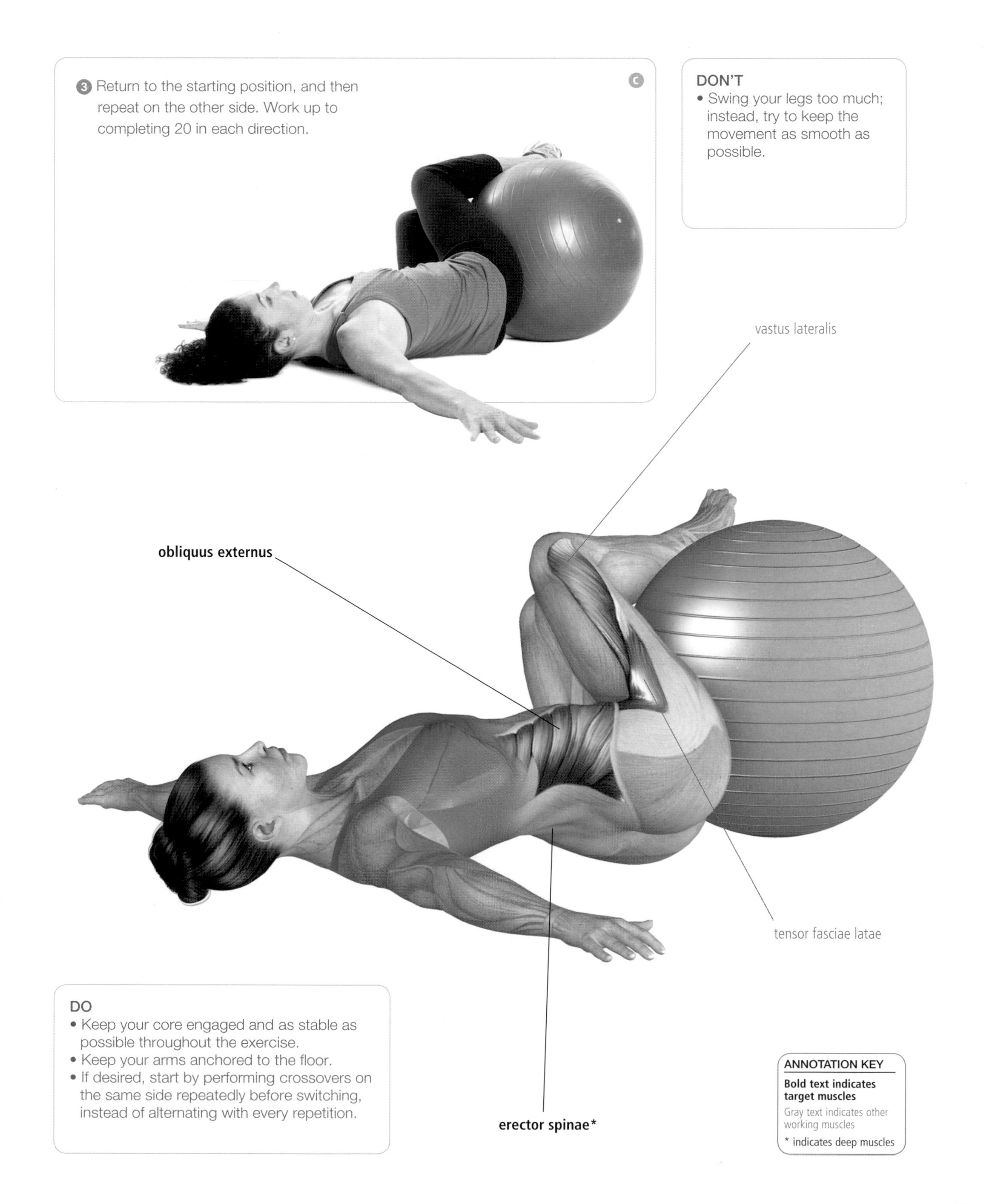

3. Return to the starting position, and then repeat on the other side. Work up to completing 20 in each direction.

DON'T
- Swing your legs too much; instead, try to keep the movement as smooth as possible.

DO
- Keep your core engaged and as stable as possible throughout the exercise.
- Keep your arms anchored to the floor.
- If desired, start by performing crossovers on the same side repeatedly before switching, instead of alternating with every repetition.

ANNOTATION KEY
Bold text indicates target muscles
Gray text indicates other working muscles
* indicates deep muscles

SEATED ARM RAISE WITH MEDICINE BALL

1. Sit on top of a Swiss ball, with your feet planted on the floor. Hold a medicine ball in your hands, and extend your arms straight out in front of you.

2. Keeping the rest of your body still, raise your arms directly over your head.

3. Lower, and repeat, performing three sets of 15.

DON'T
- Lift your feet from the floor.

DO
- Keep your torso straight and elongated.
- Gaze forward throughout the exercise.

TARGETS
- Chest
- Core
- Shoulders

LEVEL
- Beginner

BENEFITS
- Strengthens and stabilizes upper body

AVOID IF YOU HAVE . . .
- Lower-back issues
- Shoulder issues

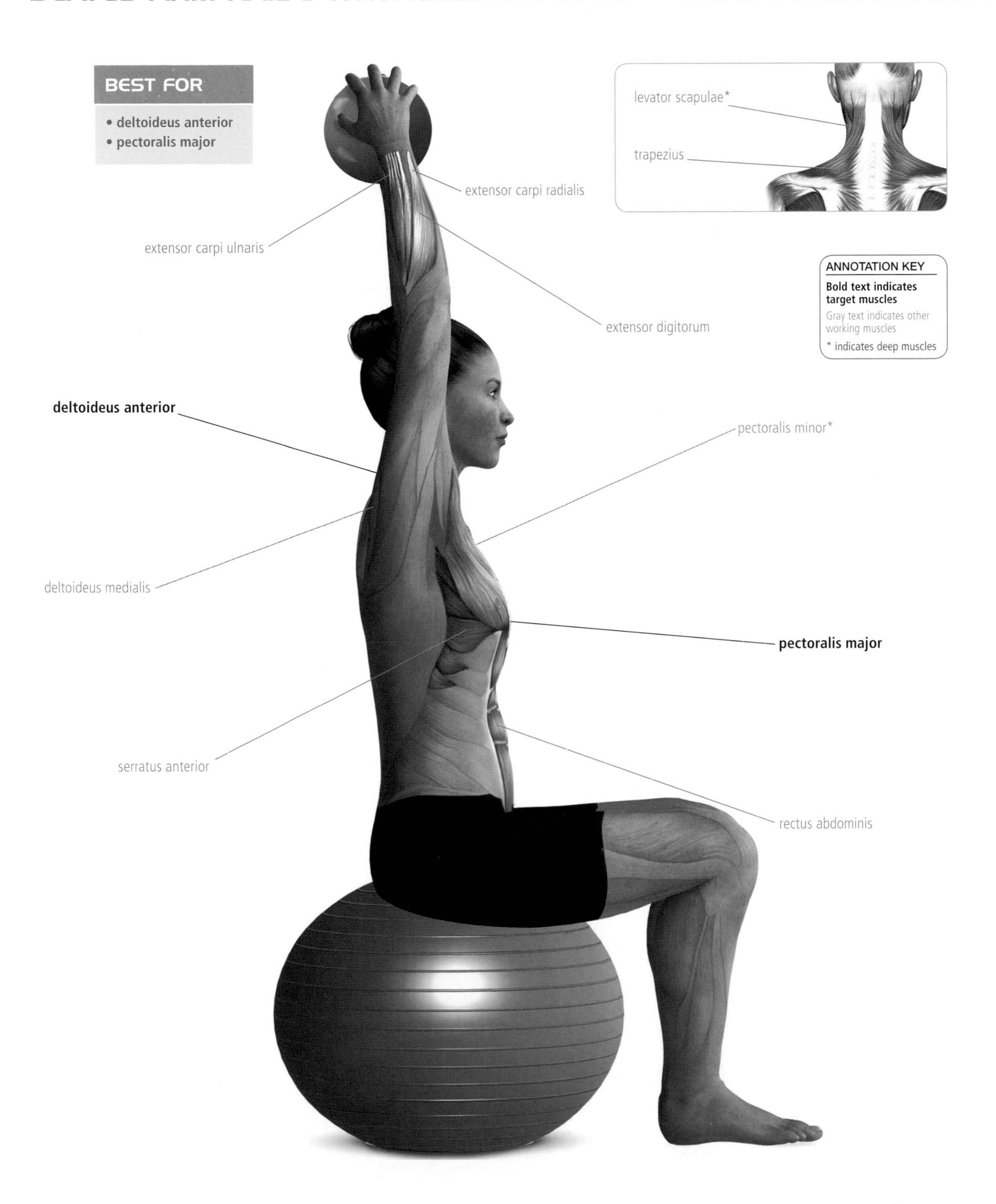
BEST FOR
• deltoideus anterior
• pectoralis major
levator scapulae*
trapezius
extensor carpi radialis
extensor carpi ulnaris
extensor digitorum
ANNOTATION KEY
Bold text indicates target muscles
Gray text indicates other working muscles
* indicates deep muscles
deltoideus anterior
pectoralis minor*
deltoideus medialis
pectoralis major
serratus anterior
rectus abdominis

MEDICINE BALL OVER-THE-SHOULDER THROW

1. Stand upright with your feet slightly more than hip-distance apart, holding a medicine ball at one side of your body. Your arms should be extended, your core rotated, and the ball held low.

a

BEST FOR

- obliquus externus
- obliquus internus
- serratus anterior
- intercostales interni
- intercostales externi

b

2. Keeping your arms extended, bring the ball in front of your body, and then continue to smoothly raise it upward to the other side as you rotate your core.

DON'T
- Move too quickly.
- Bend your arms.

TARGETS
- Chest
- Obliques

LEVEL
- Beginner

BENEFITS
- Improves core rotation
- Strengthens and stabilizes core

AVOID IF YOU HAVE . . .
- Lower-back issues

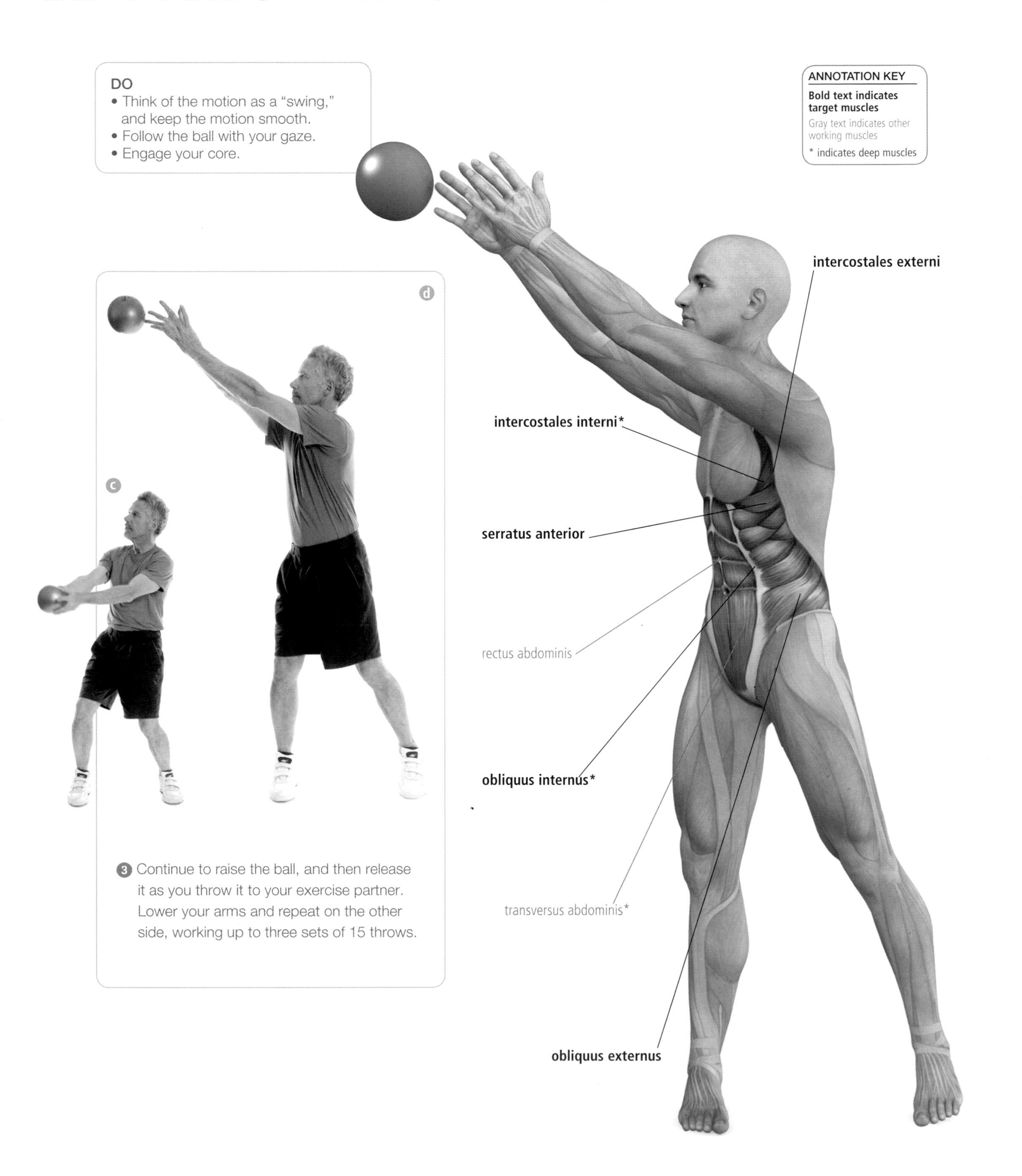

DO
- Think of the motion as a "swing," and keep the motion smooth.
- Follow the ball with your gaze.
- Engage your core.

3 Continue to raise the ball, and then release it as you throw it to your exercise partner. Lower your arms and repeat on the other side, working up to three sets of 15 throws.

BIG CIRCLES WITH MEDICINE BALL

a

1 Stand with your feet slightly farther than hip-distance apart, holding your medicine ball in front of you.

b

2 Move your arms out to one side

c

3 In a continuous circular motion, move your arms above your head.

TARGETS
- Abs
- Chest
- Obliques

LEVEL
- Beginner

BENEFITS
- Strengthens and helps to define abs
- Improves range of motion

AVOID IF YOU HAVE . . .
- Lower-back issues

BEST FOR

- rectus abdominis
- transversus abdominis
- obliquus internus
- obliquus externus
- intercostales interni
- intercostales externi

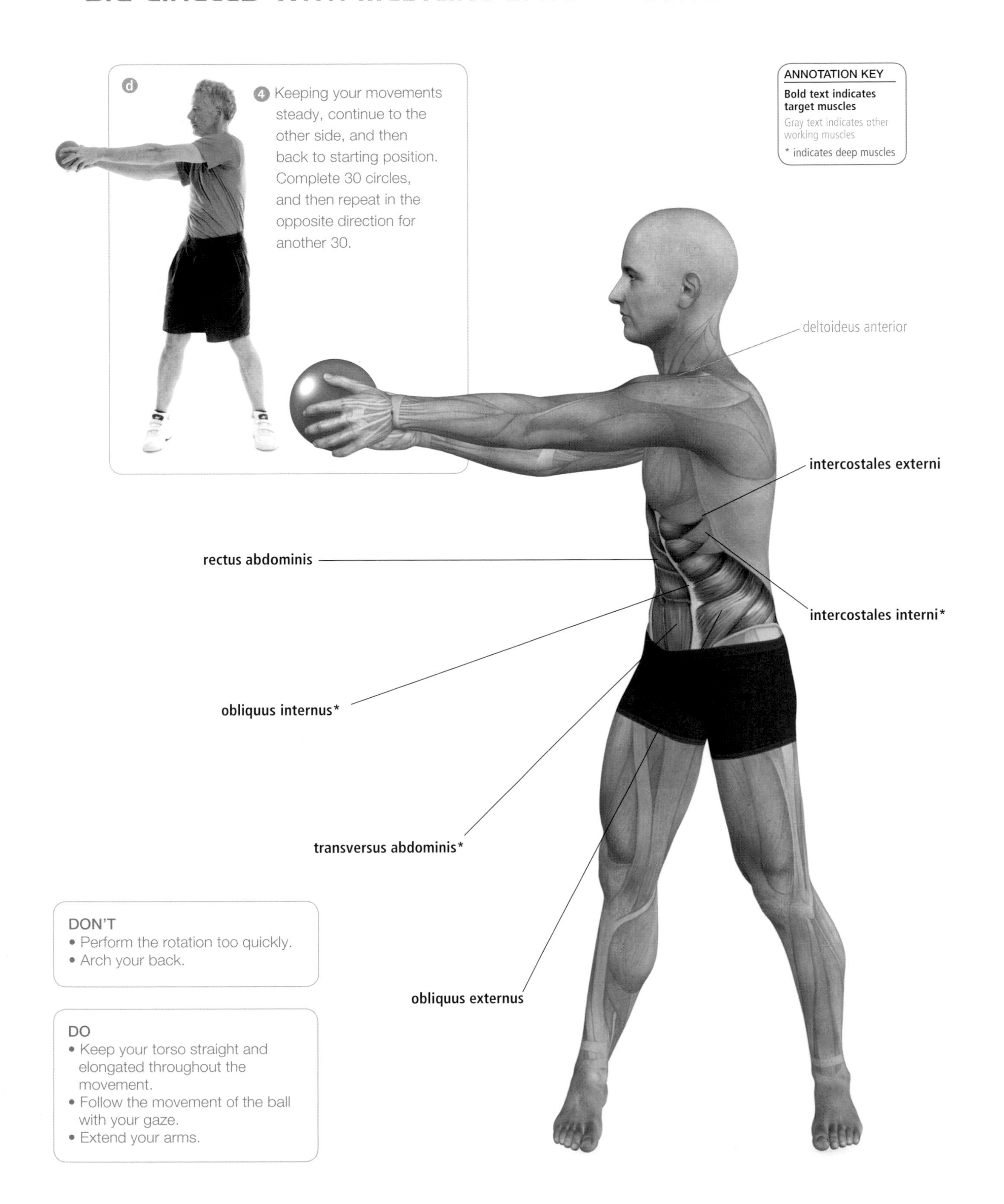

4 Keeping your movements steady, continue to the other side, and then back to starting position. Complete 30 circles, and then repeat in the opposite direction for another 30.

ANNOTATION KEY
Bold text indicates target muscles
Gray text indicates other working muscles
* indicates deep muscles

DON'T
- Perform the rotation too quickly.
- Arch your back.

DO
- Keep your torso straight and elongated throughout the movement.
- Follow the movement of the ball with your gaze.
- Extend your arms.

TONING EXERCISES

When you really want to see noticeable changes that improve both how you look and how you feel, such as dropping a dress size, losing that "gut," or taming that upper-arm jiggle, toning is the key. While cardiovascular exercises burn calories and build endurance, toning exercises firm and strengthen tissue, enhancing the ratio of muscle to fat. It's through consistently and strategically toning your muscles that you can sculpt your body into shape. The following exercises use small hand weights or adjustable dumbbells; try increasing their weight and building up the number of reps, over time.

SWISS BALL FLY

1 Lie face-up on a Swiss ball, with your upper back, neck, and head supported. Your body should be extended with your torso long, knees bent at a right angle, and feet planted on the floor a little wider than shoulder-distance apart. Grasp a hand weight or dumbbell in each hand and extend your arms straight up.

a

b

2 Keeping the rest of your body stable, bring your arms to your sides.

3 Return your arms to starting position. Repeat, completing three sets of 15.

TARGETS
- Chest

LEVEL
- Beginner

BENEFITS
- Strengthens and tones pectoral muscles

AVOID IF YOU HAVE . . .
- Shoulder issues

MODIFICATION

Same level of difficulty: Instead of holding hand weights, loop a fitness band under your ball and grasp both handles. Keep your arms extended as you hold the strap taut throughout the exercise.

a

b

DO
- When lifting the weights overhead, keep your arms directly above your shoulders.
- Keep your torso stable and feet planted throughout the exercise.
- Engage your abs.
- Keep your buttocks and pelvis lifted so that your upper legs, torso, and neck form a straight line.
- Move your arms smoothly and with control.

BEST FOR

- pectoralis major
- pectoralis minor

DON'T
- Arch your back.
- Swing your arms.

ANNOTATION KEY
Black text indicates target muscles
Gray text indicates other working muscles
* indicates deep muscles

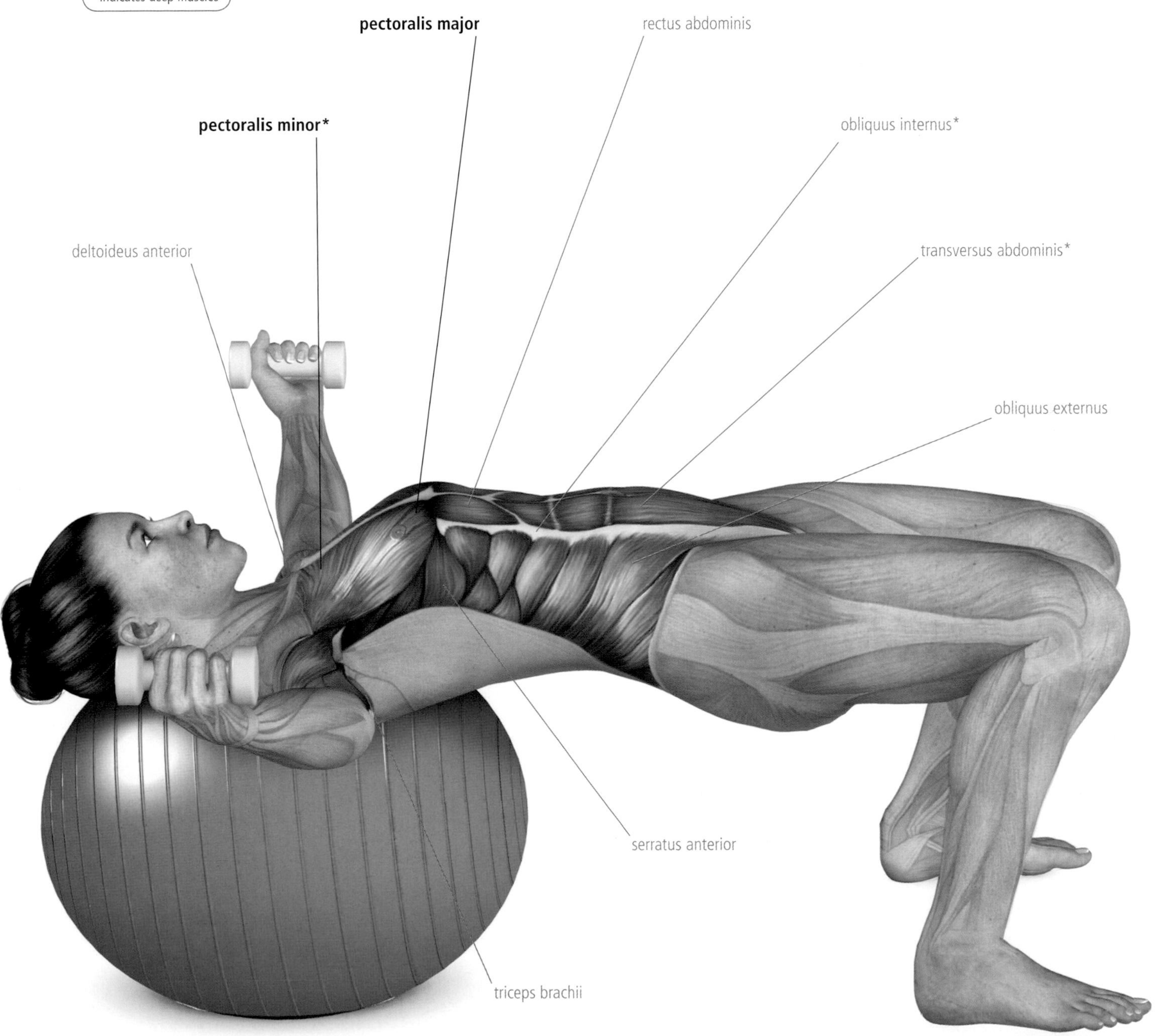

ALTERNATING FLOOR ROW

❶ Begin in completed push-up position, with a dumbbell in each hand. Your hands should be planted shoulder-width apart, palms facing each other.

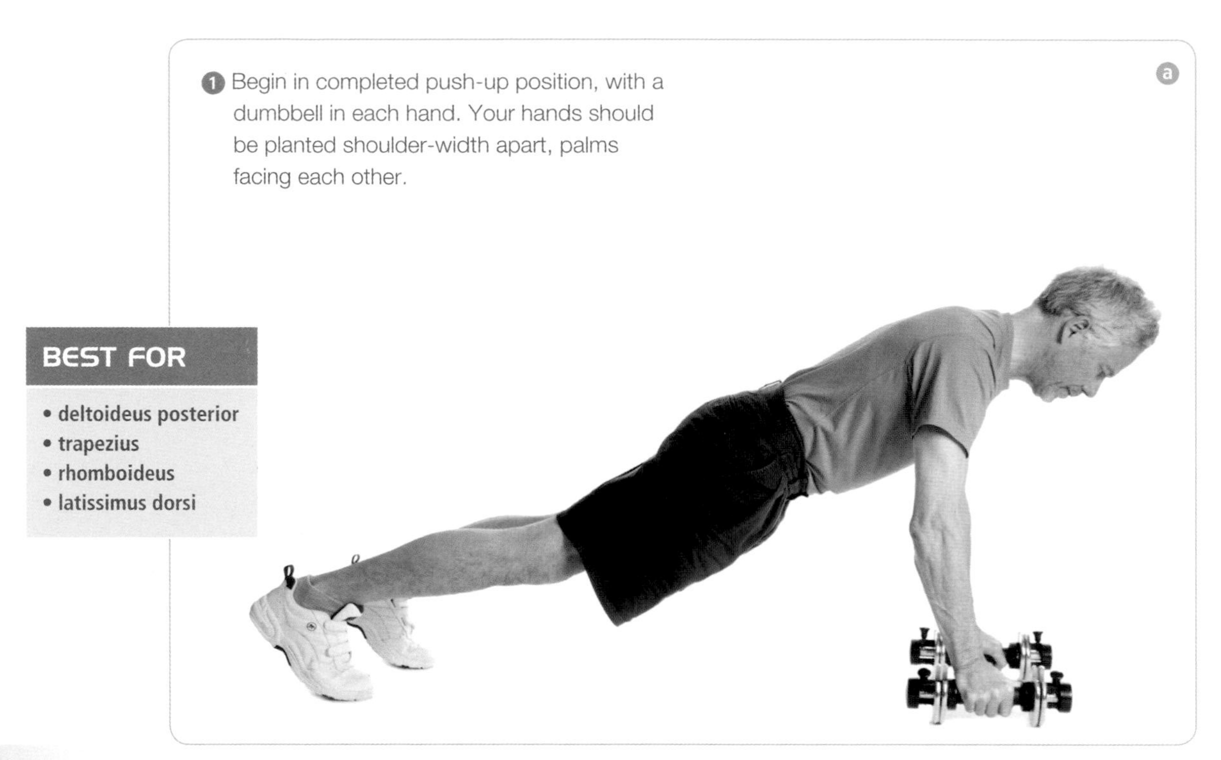

BEST FOR

- **deltoideus posterior**
- **trapezius**
- **rhomboideus**
- **latissimus dorsi**

TARGETS
- Core
- Mid back

LEVEL
- Advanced

BENEFITS
- Stabilizes core
- Strengthens middle part of back

AVOID IF YOU HAVE . . .
- Lower-back issues

❷ In a "rowing" motion, pull one dumbbell into your chest.

❸ Lower and repeat with the other arm. Perform three sets of 15 on each arm.

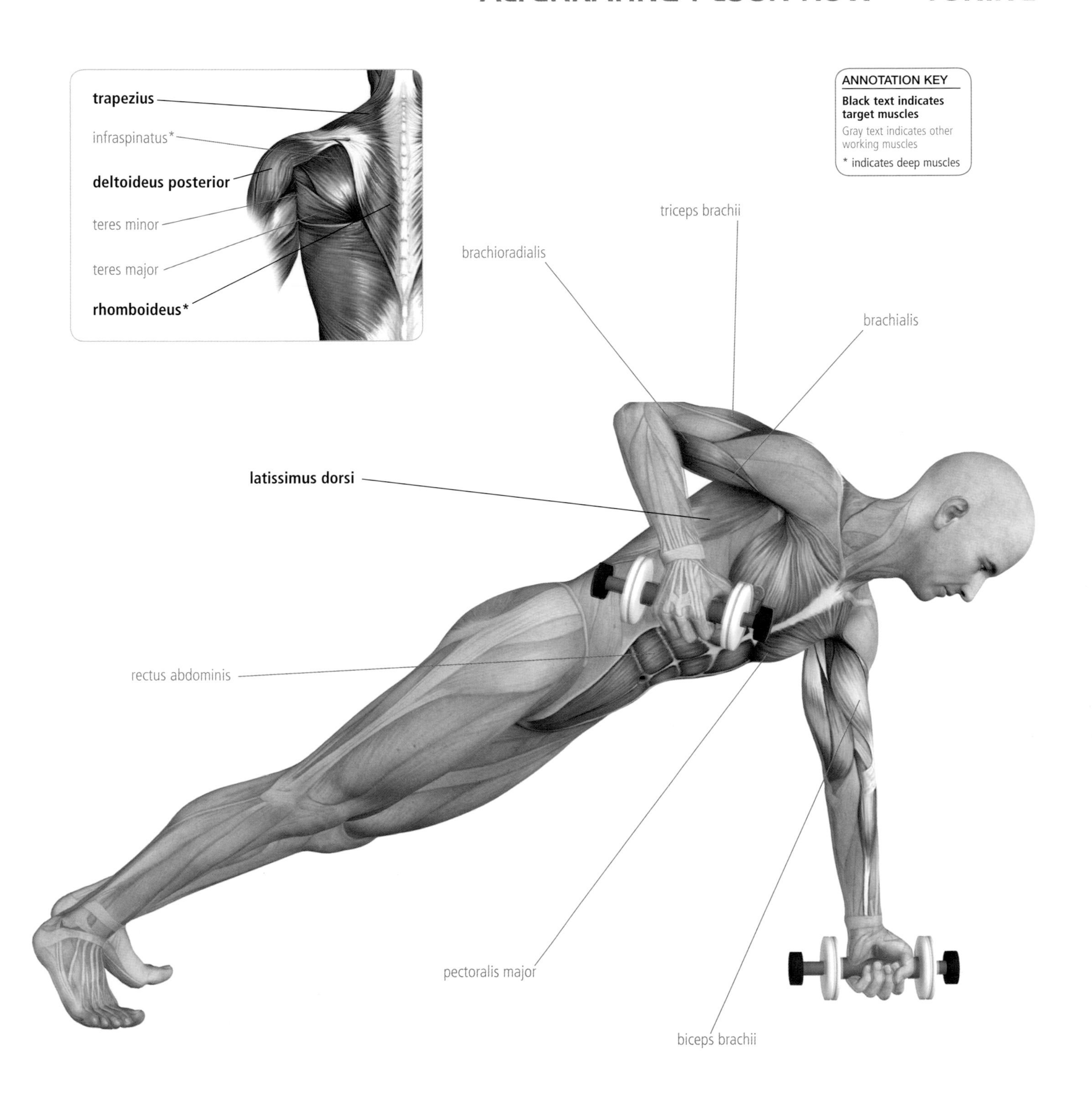

DO

- Keep your core straight.
- Move your arms smoothly and with control.
- Engage your abs and keep your torso stable.
- Keep your neck elongated and your gaze downward.

DON'T

- Rush through the exercise.
- Use momentum to drive the movement.
- Allow your lower back to sag.

SWISS BALL PULLOVER

1. Lie face-up on your Swiss ball, with your upper back, neck, and head supported. Your body should be extended with your torso long, knees bent at a right angle and feet planted on the floor a little wider than shoulder-distance apart. Grasp a hand weight or dumbbell in each hand and extend your arms behind you, level with your shoulders so that your body from knees to fingertips forms a straight line.

DON'T
- Lock your arms when they are extended behind your head.
- Arch your back.
- Rush through the exercise.

a

BEST FOR
- latissimus dorsi

TARGETS
- Core
- Upper back

LEVEL
- Intermediate

BENEFITS
- Stabilizes core
- Strengthens upper back

AVOID IF YOU HAVE . . .
- Shoulder issues

2. Keeping the rest of your body stable and your arms as straight as possible, raise your arms upward so that they are perpendicular to your body.

3. Return your arms to starting position. Repeat, performing three sets of 15.

b

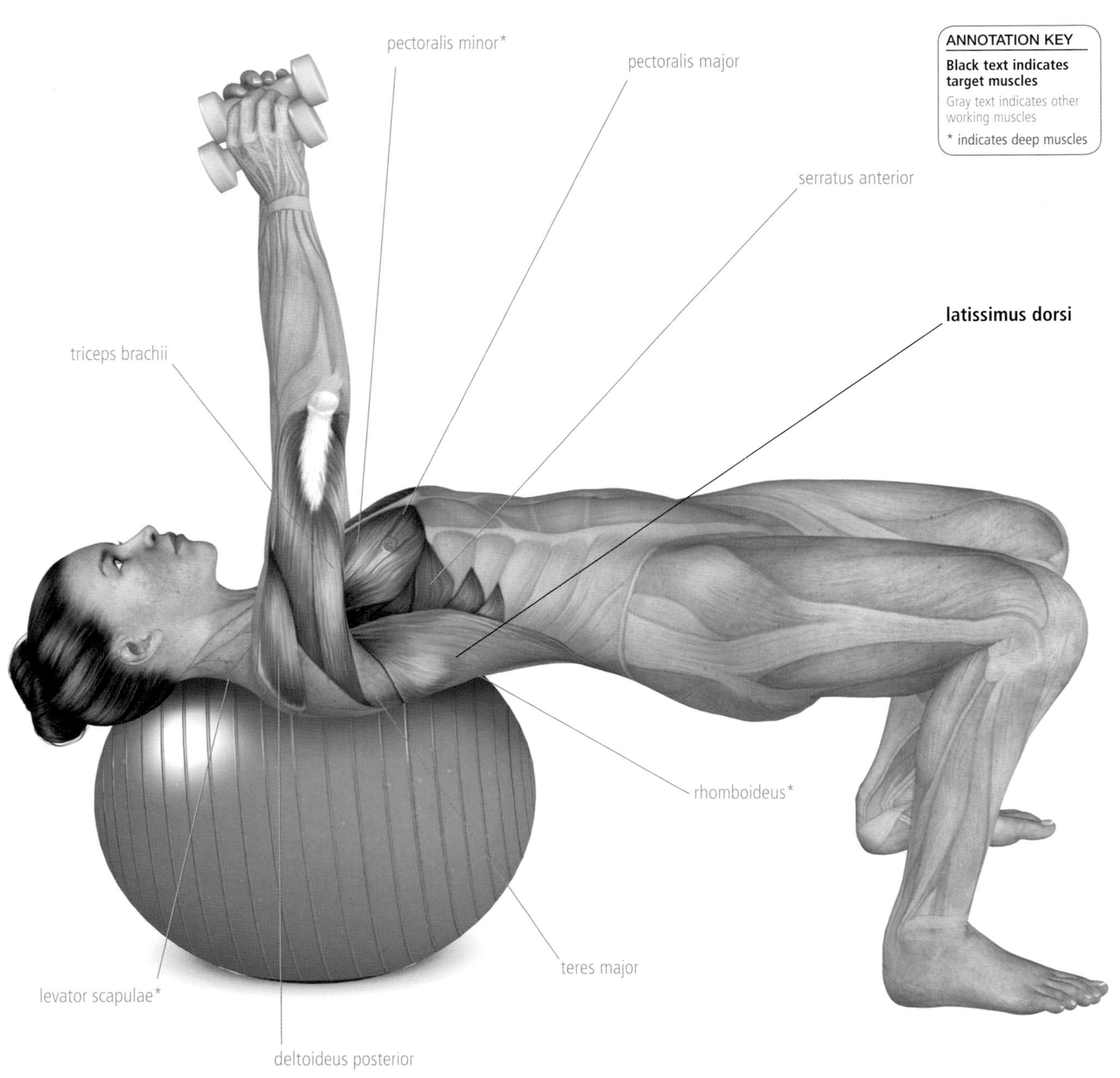

DO

- Ease into the movement.
- Keep your arms directly above your shoulders when lifting the weights overhead.
- Keep your torso stable and feet planted throughout the exercise.
- Engage your abs.
- Keep your buttocks and pelvis lifted so that your upper legs, torso, and neck form a straight line.
- Move your arms smoothly and with control.

MODIFICATION

Same level of difficulty: Instead of using hand weights, grasp a medicine ball in your hands as you perform the exercise.

SWISS BALL SEATED SHOULDER PRESS

a

1. Sit upright on a Swiss ball, with your feet planted a little wider than shoulder-width apart. Grasp a hand weight or dumbbell in each hand, bend your elbows, and raise your arms so that your forearms are parallel to the floor and the dumbbells are level with your head.

BEST FOR

- **deltoideus anterior**

b

2. Extend your arms upward, raising the dumbbells over your head.
3. Return your arms to starting position and repeat. Perform three sets of 15.

TARGETS
- Shoulders

LEVEL
- Beginner

BENEFITS
- Strengthens deltoids

AVOID IF YOU HAVE . . .
- Lower-back issues
- Shoulder issues

DO
- Keep your palms facing forward throughout the exercise.
- Try to extend your arms all the way upward.
- Move slowly and with control.
- Gaze forward.
- Keep your abs engaged and your core stable.

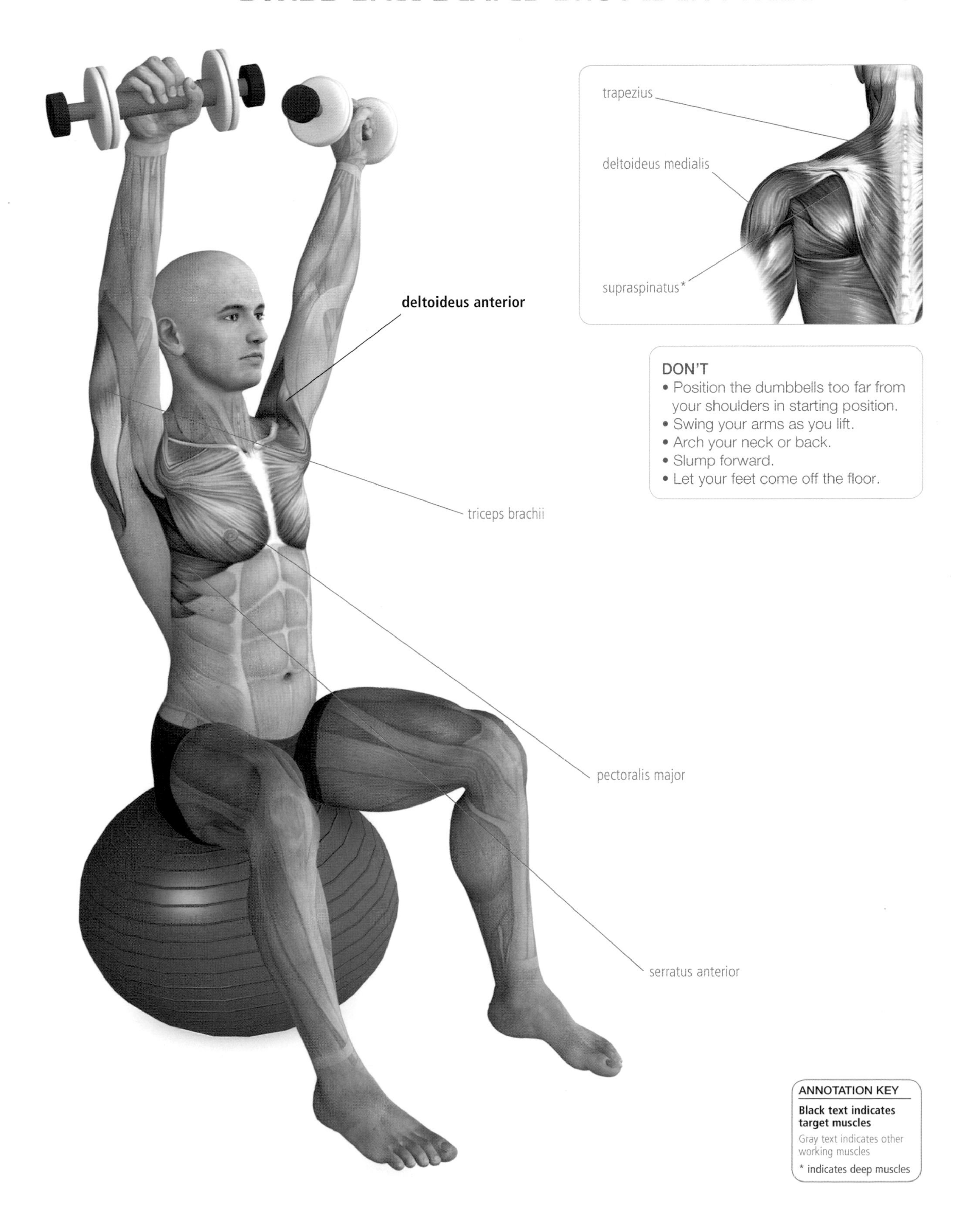

DON'T

- Position the dumbbells too far from your shoulders in starting position.
- Swing your arms as you lift.
- Arch your neck or back.
- Slump forward.
- Let your feet come off the floor.

ANNOTATION KEY

Black text indicates target muscles

Gray text indicates other working muscles

* indicates deep muscles

ALTERNATING DUMBBELL CURL

1. Stand upright, with your feet planted about shoulder-width apart and your knees very slightly bent. Hold a hand weight or dumbbell in each hand, with your arms down along your sides, palms facing forward.

a

BEST FOR

- biceps brachii

2. In a smooth, controlled movement, bend one arm as you raise the weight toward your shoulder.

3. As you begin to lower your arm, begin to raise the other one, and repeat on the other side. Continue to alternate, completing three sets of 15 per arm.

b

TARGETS
- Biceps

LEVEL
- Beginner

BENEFITS
- Strengthens and tones biceps

AVOID IF YOU HAVE . . .
- Lower-back issues

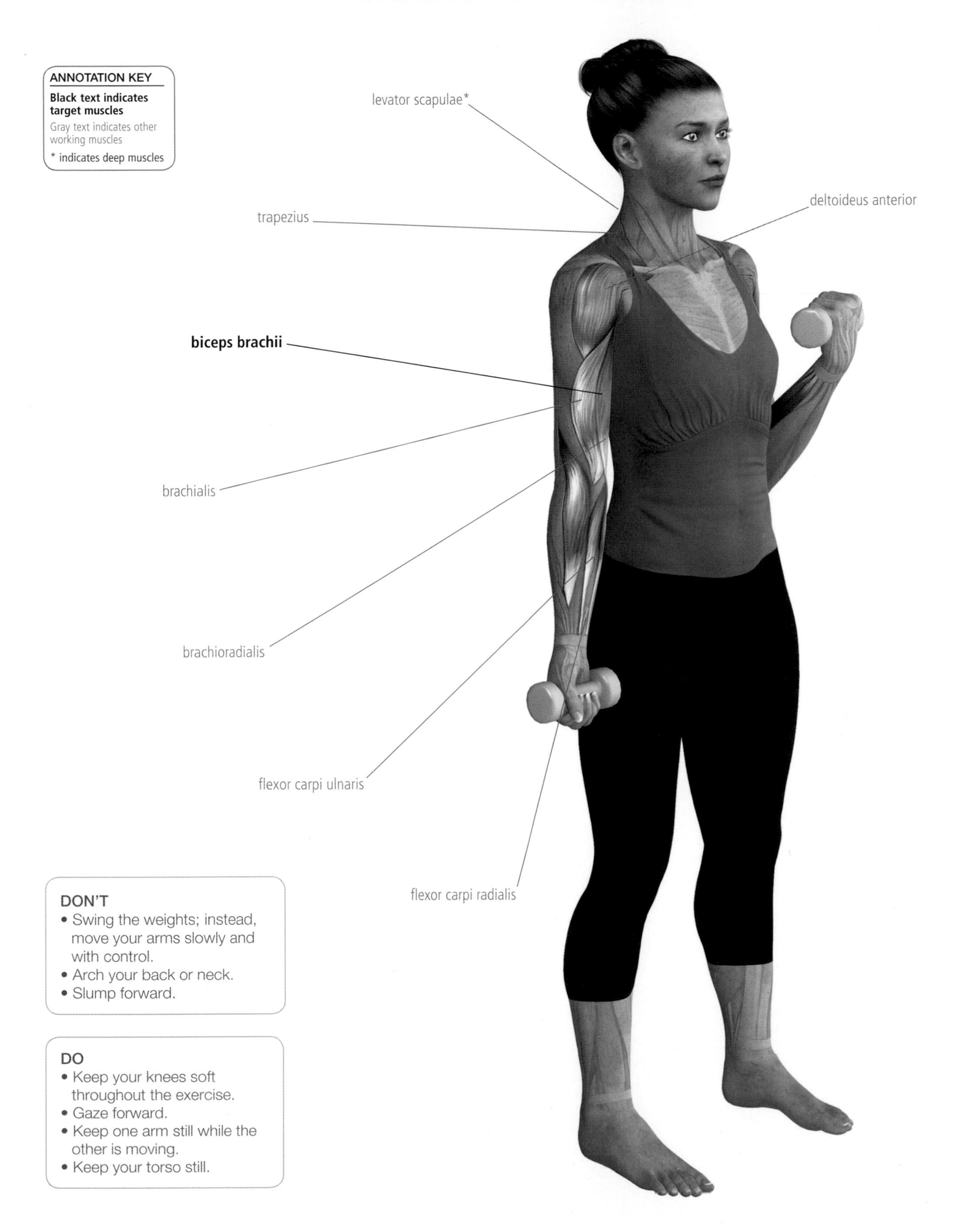

DON'T

- Swing the weights; instead, move your arms slowly and with control.
- Arch your back or neck.
- Slump forward.

DO

- Keep your knees soft throughout the exercise.
- Gaze forward.
- Keep one arm still while the other is moving.
- Keep your torso still.

LYING TRICEPS EXTENSION

1. Lie face-up on a Swiss ball, with your upper back, neck, and head supported. Your body should be extended with your torso long, knees bent at a right angle and feet planted on the floor a little wider than shoulder-distance apart. Grasp a hand weight or dumbbell in each hand and extend your arms straight up.

a

BEST FOR
- triceps brachii

TARGETS
- Triceps

LEVEL
- Intermediate

BENEFITS
- Strengthens and tones triceps

AVOID IF YOU HAVE . . .
- Elbow pain

2. Bend your elbows as you lower the weights toward your head.

3. Straighten your arms upward to starting position and then repeat. Perform three sets of 15 repetitions.

b

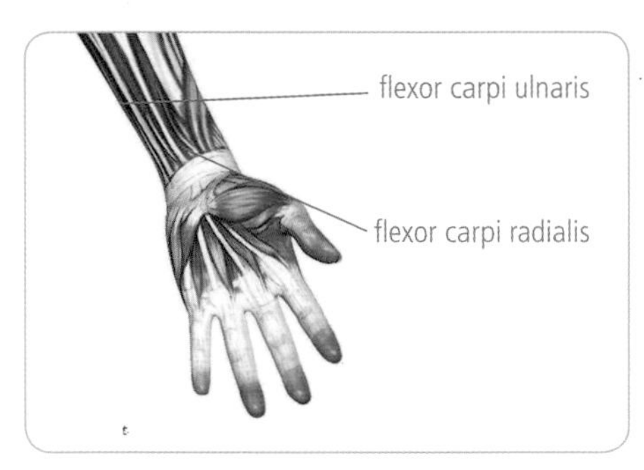

DO
- Keep your forearms stable and your elbows over your shoulders.
- Keep your torso stable and feet planted throughout the exercise.
- Engage your abs.
- Keep your glutes and pelvis lifted so that your upper legs, torso, and neck form a straight line.
- Move smoothly and with control.

DON'T
- Arch your back.
- Flare your elbows outward.
- Swing your weights—especially important as the weights are close to your head.

ANNOTATION KEY
Black text indicates target muscles
Gray text indicates other working muscles
* indicates deep muscles

SUMO SQUAT

1. Stand with your feet apart and turned out, holding a dumbbell between your legs.

DO
- Gaze forward.
- Keep your chest lifted and your shoulders down.
- Engage your core.

DON'T
- Allow your knees to extend past your feet.
- Arch your back or slump forward.
- Hunch your shoulders.
- Twist your torso.

TARGETS
- Buttocks
- Fronts of thighs

LEVEL
- Beginner

BENEFITS
- Tones glutes and thighs

AVOID IF YOU HAVE . . .
- Lower-back issues

2. Keeping your torso upright, bend your knees as you lower into a squat position.

3. Push through your heels as you rise back into an upright position. Repeat, completing three sets of 15.

BEST FOR
- gluteus maximus

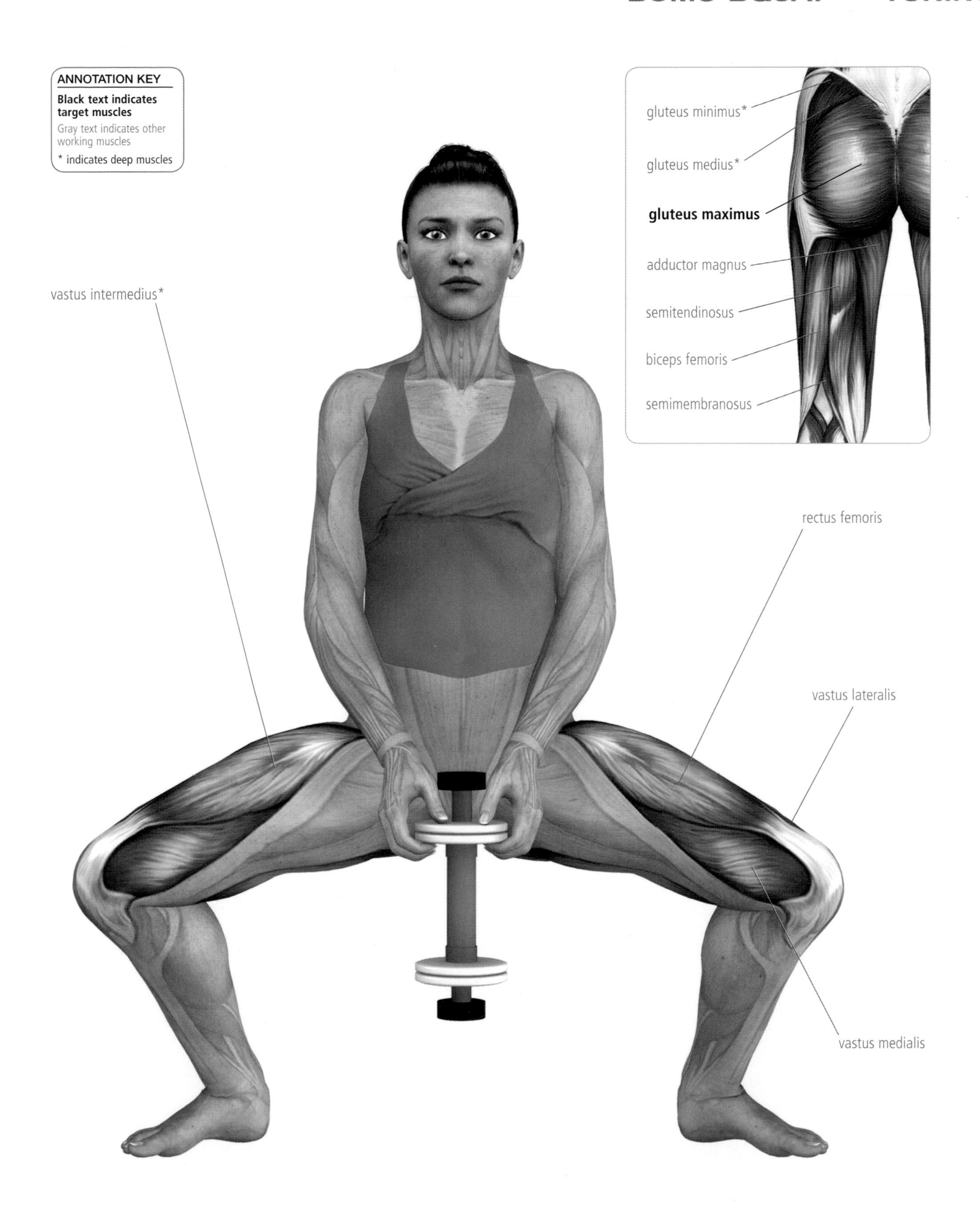
ANNOTATION KEY
Black text indicates target muscles
Gray text indicates other working muscles
* indicates deep muscles
gluteus minimus*
gluteus medius*
gluteus maximus
adductor magnus
semitendinosus
biceps femoris
semimembranosus
vastus intermedius*
rectus femoris
vastus lateralis
vastus medialis

LUNGE

❶ Stand upright, feet planted about shoulder-width apart, with your arms at your sides and a hand weight or dumbbell in each hand.

DO
- Keep your body facing forward as you step one leg in front of you.
- Stand upright.
- Gaze forward.
- Ease into the lunge.
- Make sure that your front knee is facing forward.

TARGETS
- Buttocks
- Fronts of thighs

LEVEL
- Intermediate

BENEFITS
- Strengthens and tones quadriceps and glutes

AVOID IF YOU HAVE . . .
- Knee issues

BEST FOR
- **gluteus maximus**
- **rectus femoris**
- **vastus lateralis**
- **vastus intermedius**
- **vastus medialis**

❷ Keeping your head up and your spine neutral, take a big step forward.

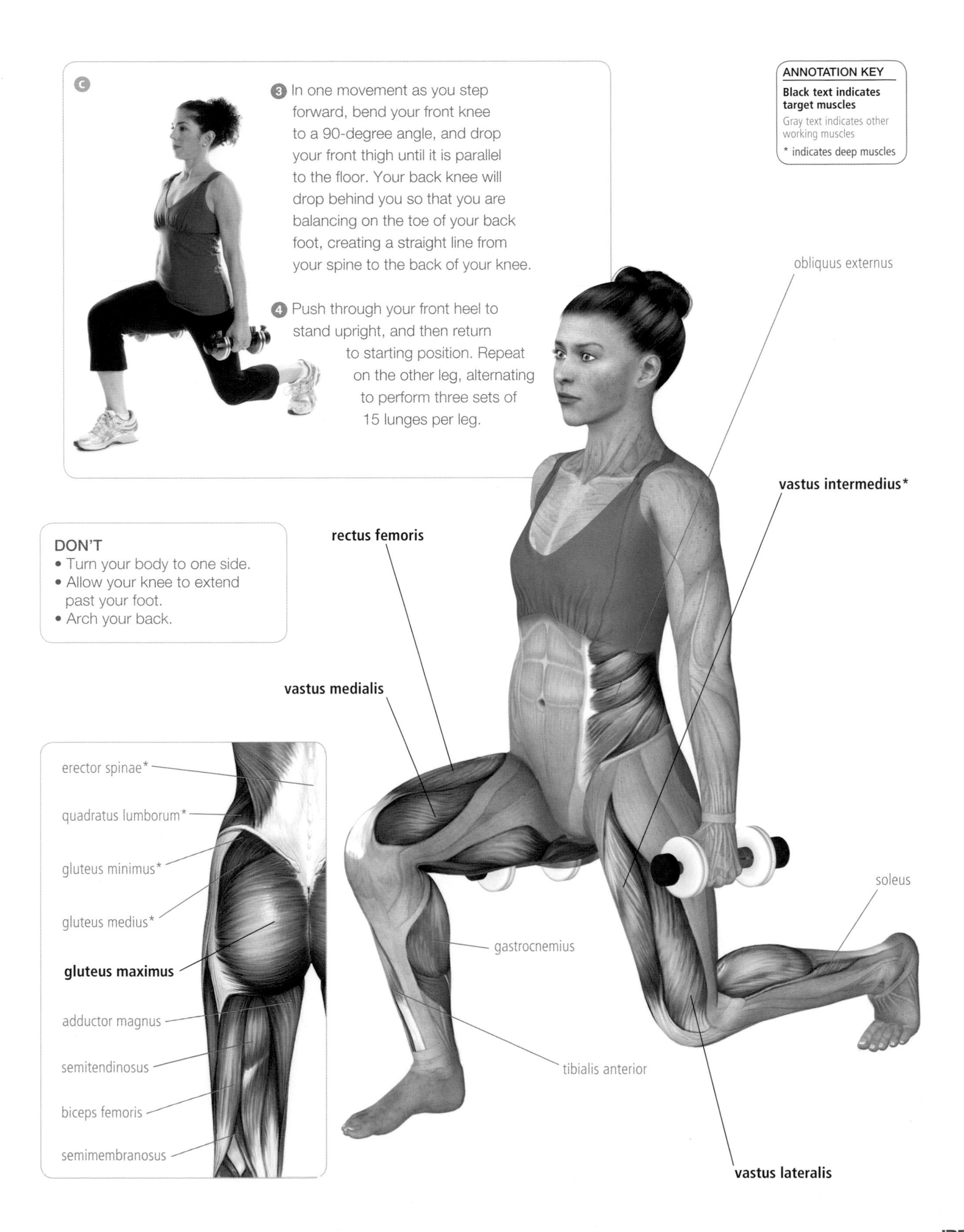

3 In one movement as you step forward, bend your front knee to a 90-degree angle, and drop your front thigh until it is parallel to the floor. Your back knee will drop behind you so that you are balancing on the toe of your back foot, creating a straight line from your spine to the back of your knee.

4 Push through your front heel to stand upright, and then return to starting position. Repeat on the other leg, alternating to perform three sets of 15 lunges per leg.

DON'T
- Turn your body to one side.
- Allow your knee to extend past your foot.
- Arch your back.

STIFF-LEGGED DEADLIFT

a

1. Stand upright, feet planted about shoulder-width apart, with your arms slightly in front of your thighs with a hand weight or dumbbell in each hand. Your knees should be slightly bent and your rear pushed slightly outward.

2. Keeping your back flat, hinge at the hips and bend forward as you lower the dumbbells toward the floor. You should feel a stretch in the backs of your legs.

3. With control, raise your upper body back to starting position. Repeat, completing three sets of 15.

b

TARGETS
- Back
- Buttocks
- Hamstrings

LEVEL
- Intermediate

BENEFITS
- Improves flexibility and stabilization throughout lower body

AVOID IF YOU HAVE . . .
- Lower-back issues

DO
- Maintain the straight line of your back.
- Keep your torso stable.
- Keep your neck straight.
- Keep your arms extended.

DON'T
- Allow your lower back to sag or arch.
- Arch your neck, straining to look forward while you are bent over.

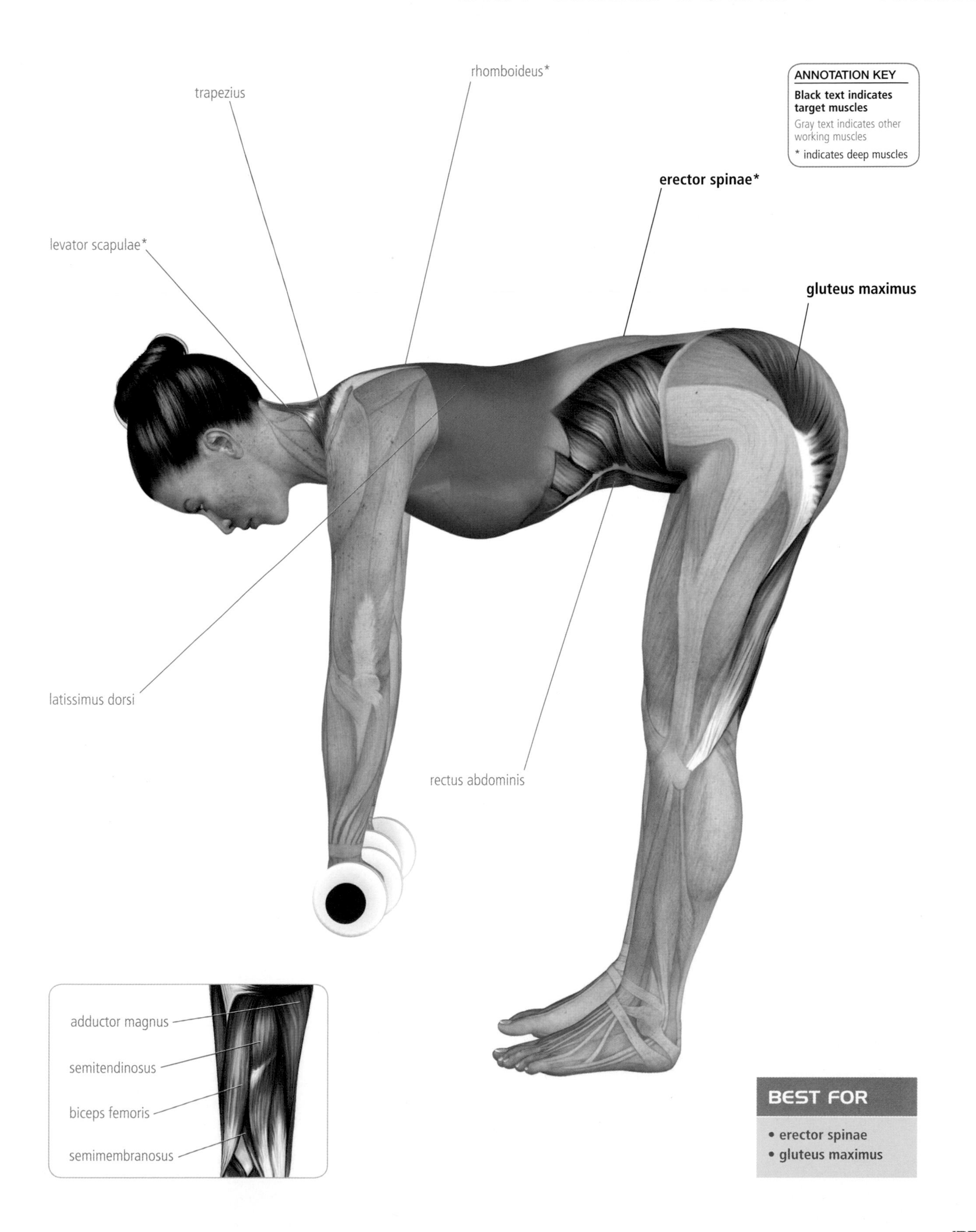

BEST FOR

- erector spinae
- gluteus maximus

DUMBBELL CALF RAISE

❶ Stand with your arms at your sides, holding a hand weight or dumbbell in each hand with palms facing inward.

ⓐ

BEST FOR

- gastrocnemius

❷ Keeping the rest of your body steady, slowly raise your heels off the floor to balance on the balls of your feet.

❸ Hold for 10 seconds, lower, and repeat, performing three sets of 15.

ⓑ

TARGETS
- Calves

LEVEL
- Intermediate

BENEFITS
- Strengthens calf muscles

AVOID IF YOU HAVE . . .
- Ankle issues

DO
- Keep your legs straight.
- Concentrate on the contraction in your calves as you balance on the balls of your feet; to feel a greater contraction, rise higher.
- Keep your core stable and your back straight.
- Gaze forward.
- Try to balance on the balls of your feet.

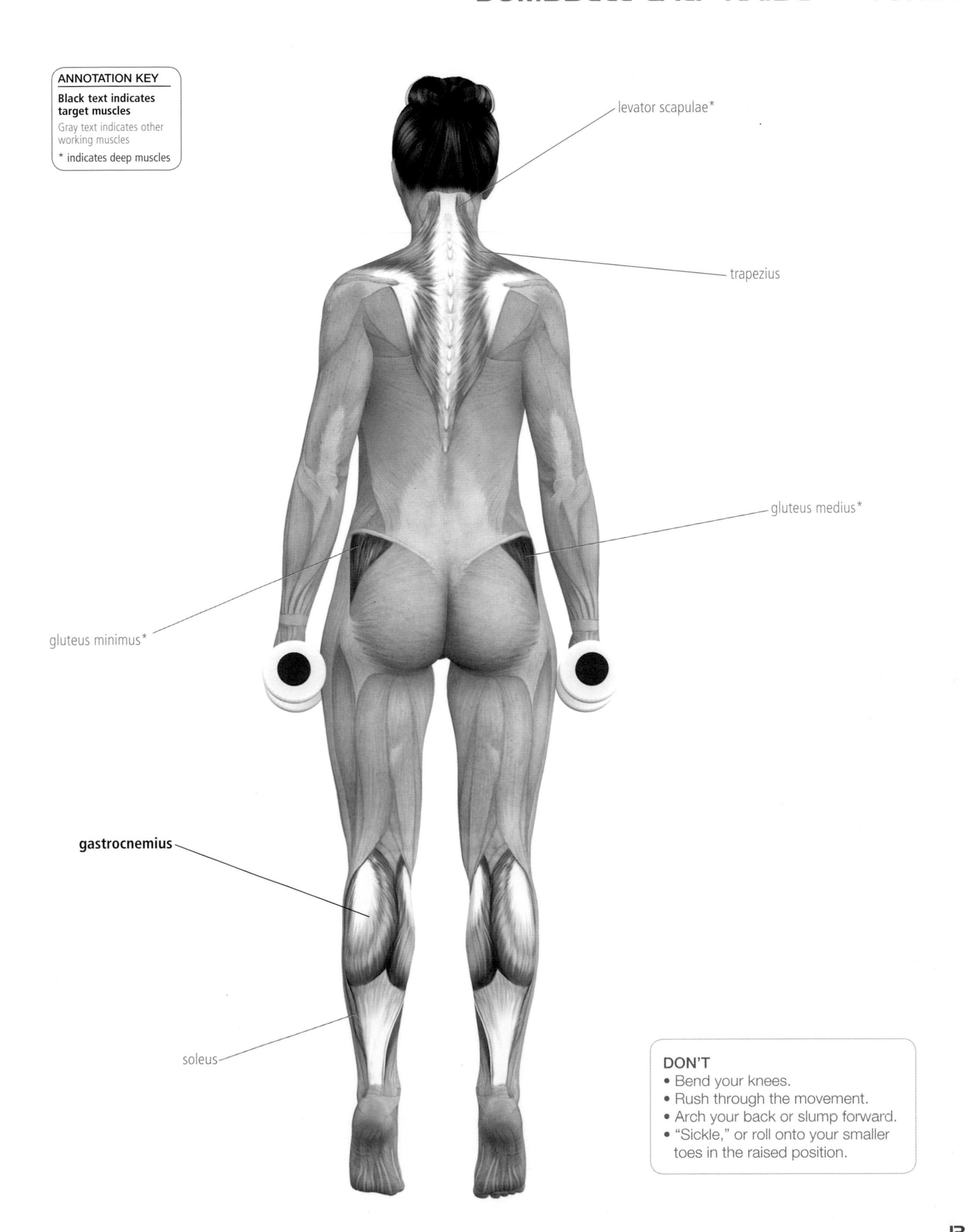

DON'T
- Bend your knees.
- Rush through the movement.
- Arch your back or slump forward.
- "Sickle," or roll onto your smaller toes in the raised position.

SWISS BALL INCLINE CHEST PRESS

TONING

1 Lie face-up on a Swiss ball, with your upper back, neck, and head supported. Your body should be extended with your torso long, knees bent at a right angle and feet planted on the floor a little wider than shoulder-distance apart. Grasp a hand weight or dumbbell in each hand and extend your arms straight up.

a

BEST FOR

- pectoralis major
- pectoralis minor

b

2 Drop your rear slightly so that your torso takes on an "incline" position.

3 Bend your elbows as you lower the weights toward your shoulders.

4 Extend your arms upward to starting position. Repeat, performing three sets of 15.

TARGETS
- Chest
- Core

LEVEL
- Intermediate

BENEFITS
- Strengthens pectoral muscles
- Strengthens and stabilizes core

AVOID IF YOU HAVE . . .
- Shoulder issues
- Lower-back issues

DON'T
- Rush through the movement.
- Twist to one side.
- Move your torso as you lift or lower your arms.

DO
- Synchronize the movement of your arms.
- Keep your abs engaged, your core stable, and your pelvis squared as you maintain the incline position.
- Move your arms slowly and with control.

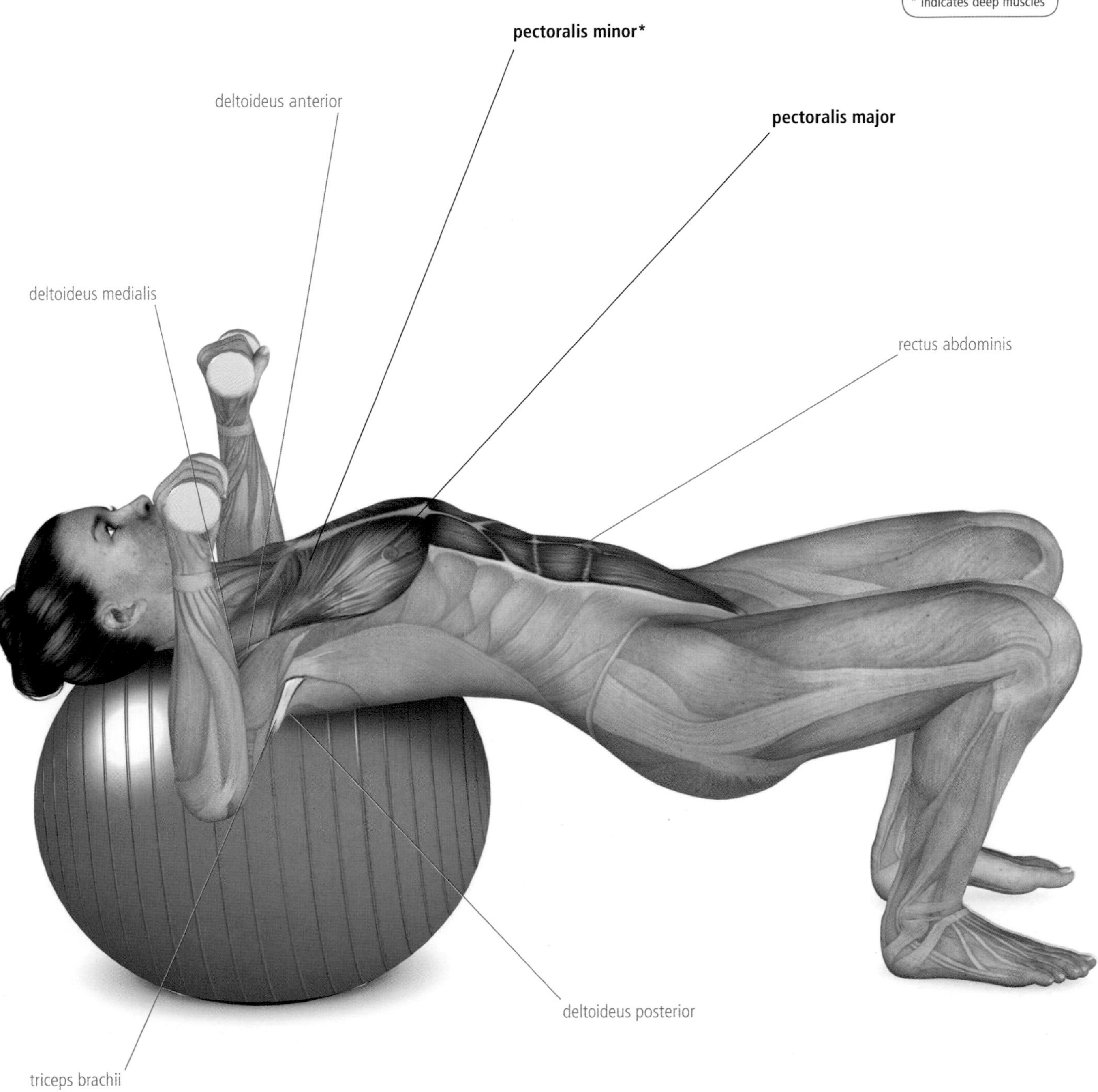
ANNOTATION KEY
Black text indicates target muscles
Gray text indicates other working muscles
* indicates deep muscles
pectoralis minor*
deltoideus anterior
pectoralis major
deltoideus medialis
rectus abdominis
deltoideus posterior
triceps brachii

DUMBBELL UPRIGHT ROW

a

1. Stand holding a pair of dumbbells in front of your thighs.

2. Bend your elbows to the side as you raise your weights, aiming for shoulder height.
3. Lower the dumbbells to starting position. Repeat, completing three sets of 15.

b

TARGETS
- Shoulders
- Upper back

LEVEL
- Beginner

BENEFITS
- Strengthens muscles in upper back and shoulders

AVOID IF YOU HAVE . . .
- Shoulder issues
- Tennis elbow

DON'T
- Swing your weights; instead, move slowly and with control.
- Arch your back or slump forward.

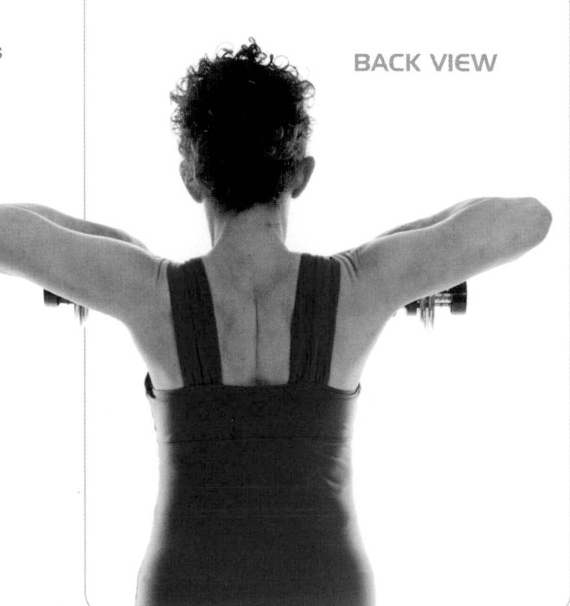

BACK VIEW

DO
- Keep your torso stable, your back straight, and your abs engaged.
- Lead with your elbows.

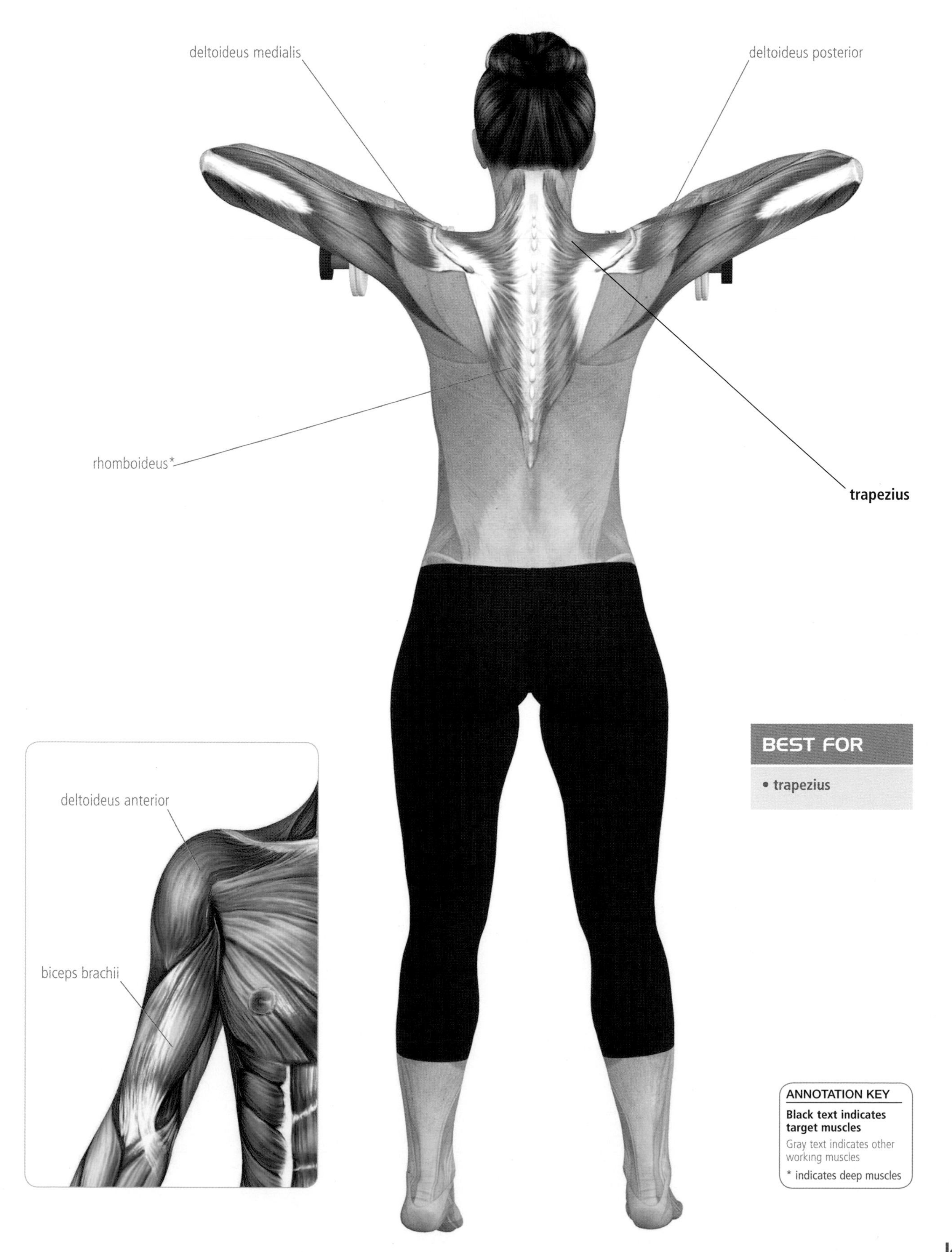

BEST FOR

- trapezius

ANNOTATION KEY

Black text indicates target muscles

Gray text indicates other working muscles

* indicates deep muscles

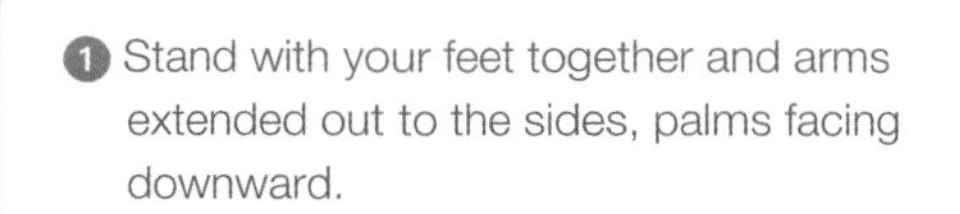

SIDE LUNGE

❶ Stand with your feet together and arms extended out to the sides, palms facing downward.

❷ In one smooth movement, step one foot out to the side as you bend the opposite knee. As your chest moves forward and your hips move back, extend your arms to the front to maintain your balance.

❸ Pushing off the stepping leg, bring your arms and legs back to the starting position, and repeat on the other side. Perform three sets of 15 per side.

DO
- Move your arms and hips at the same time.
- Keep your chest up and your shoulders down.
- Keep your upper arms parallel to the floor.
- Engage your glutes as you lunge.

TARGETS
- Glutes
- Quads

LEVEL
- Intermediate

BENEFITS
- Strengthens and tones glutes and quads
- Improves balance

AVOID IF YOU HAVE . . .
- Knee issues

BEST FOR
- gluteus maximus
- rectus femoris
- vastus lateralis
- vastus intermedius
- vastus medialis

DON'T
- Rush through the exercise.
- Arch your back or slump forward.

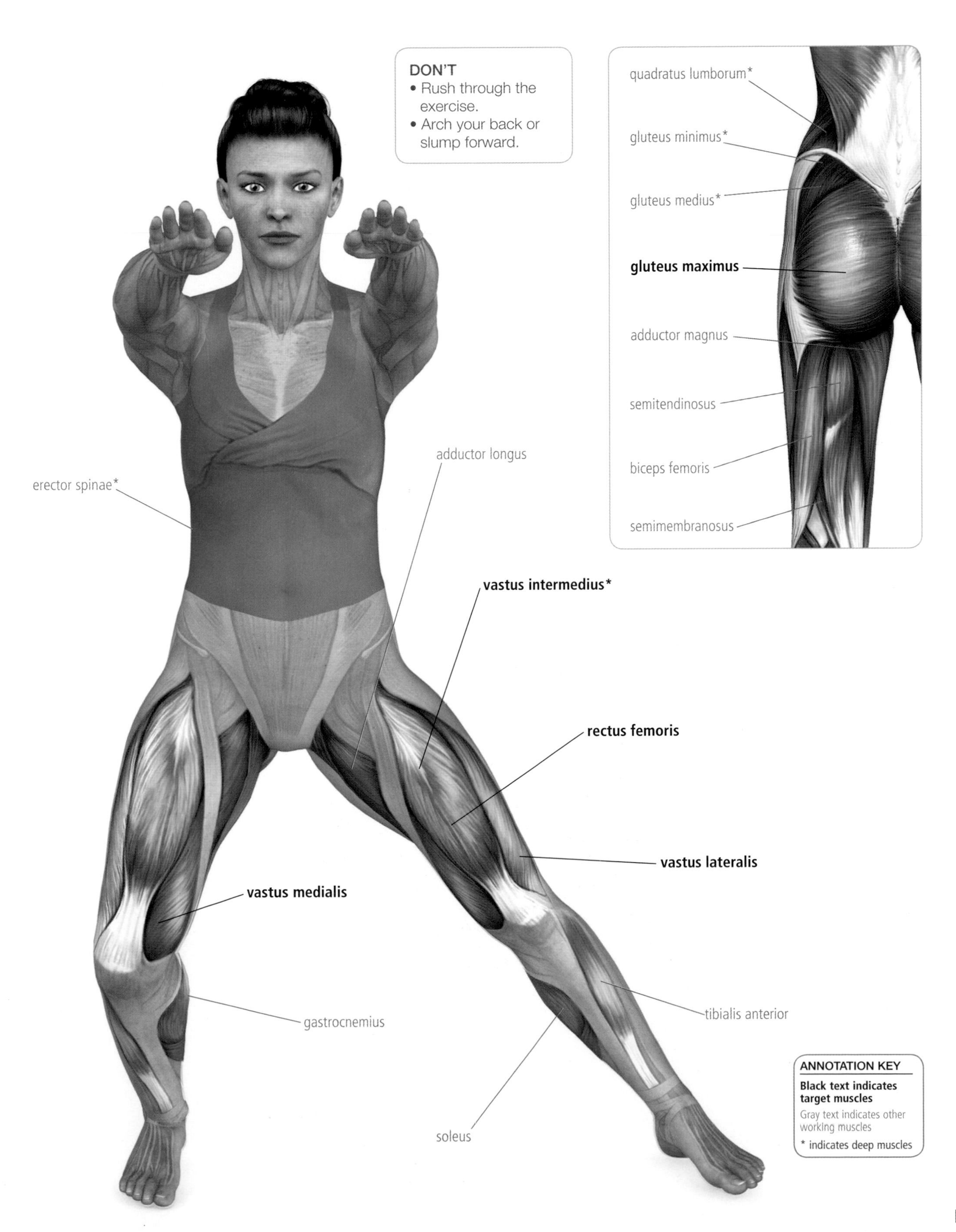

ANNOTATION KEY
Black text indicates target muscles
Gray text indicates other working muscles
* indicates deep muscles

SWISS BALL HAMSTRINGS CURL

❶ Lie on your back with your arms along your sides, angled slightly away from your body. Extend your legs and rest your lower legs and ankles on top of a Swiss Ball.

BEST FOR

- biceps femoris
- semitendinosus
- semimembranosus

a

DO

- Position your legs on the ball to form a 45-degree angle with the rest of your body before you curl.
- Move smoothly, maintaining control of the ball.
- Keep your arms anchored to the floor.
- Engage your abs, and squeeze your glutes.

TARGETS

- Glutes
- Hamstrings

LEVEL

- Intermediate

BENEFITS

- Strengthens and tones hamstrings and glutes

AVOID IF YOU HAVE . . .

- Lower-back issues
- Shoulder issues
- Neck issues

❷ Pressing downward with your feet, bend your knees as you roll the ball toward you. Curl your pelvis, and raise your lower body off the floor. Hold for 5 seconds.

❸ With control, return to starting position and then repeat, working up to three sets of 15 repetitions.

b

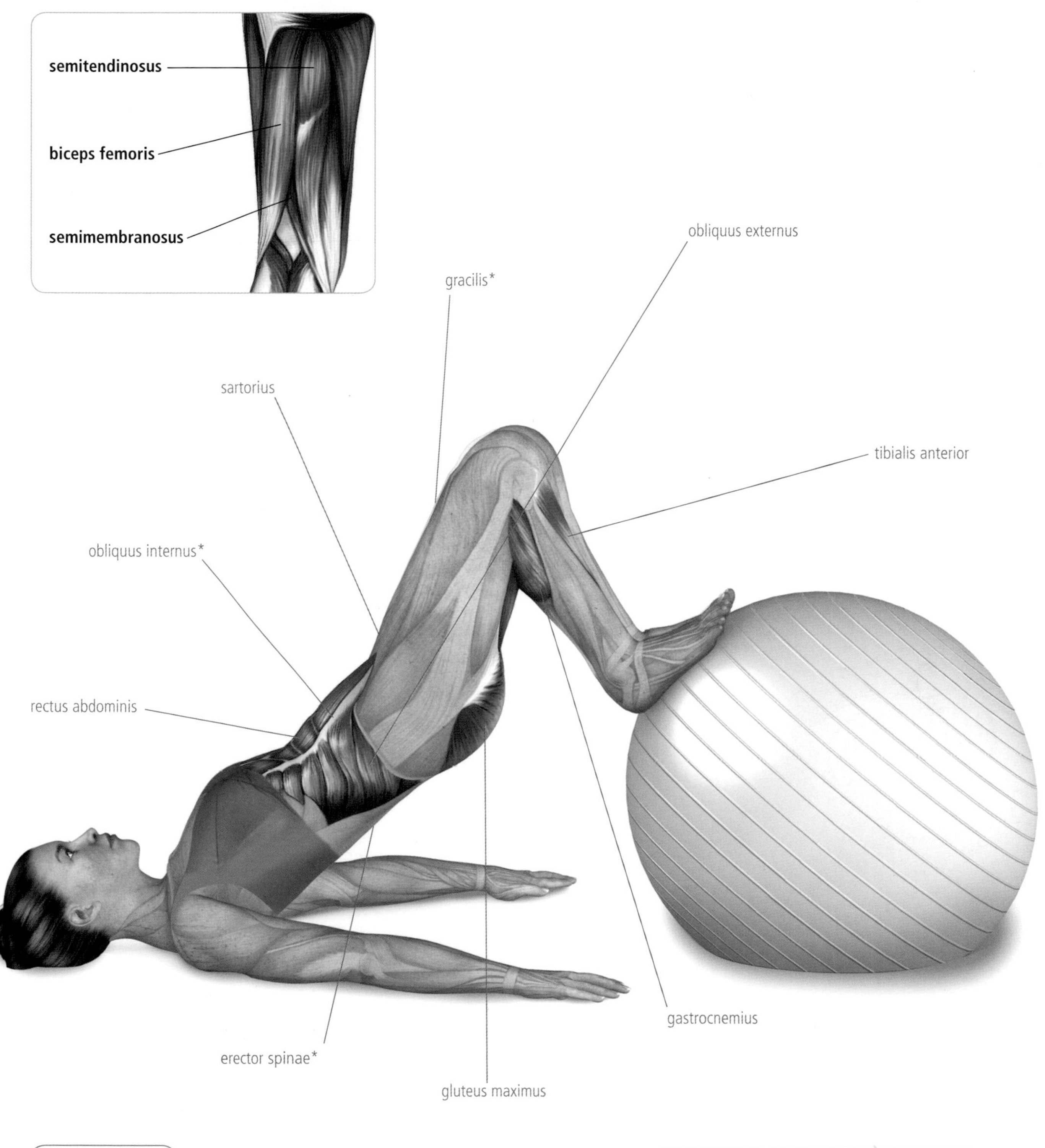

ANNOTATION KEY

Black text indicates target muscles

Gray text indicates other working muscles

* indicates deep muscles

DON'T

- Rush through the movement.
- Arch your back in the curl position; instead, keep it as straight as possible.

WORKOUTS

Once you have gone through the exercises in this book and practiced executing them properly, your next step is to put these moves together into workouts. The following pages give you several sample workouts designed with particular aims in mind, whether you are a beginner just getting used to exercising or an advanced exerciser who wants to craft a workout that achieves a certain goal, such as improving your posture or toning your legs and hips. After trying the workouts featured here, try flipping through the exercises in this book and create your own workouts to suit your individual fitness goals.

BEGINNER'S WORKOUT

Although it is especially beneficial if you are new to training, this all-around workout is suitable for all levels.

❶ Swiss Ball Fly

pages 120–121

❷ Swiss Ball Pullover

pages 124–125

❸ Alternating Dumbbell Curl

pages 128–129

❹ Lying Triceps Extension

pages 130–131

❺ Sumo Squat

pages 132–133

❻ Seated Arm Raise with Medicine Ball

pages 112–113

❼ Crunch

pages 100–101

❽ Leg Raise

pages 104–105

❾ Seated Russian Twist

pages 106–107

❿ Bridge

pages 60–61

⓫ Prone Arm Raise

page 62

POSTURE WORKOUT

This combination of exercises strengthens and supports the postural muscles.

❶ **Swiss Ball Seated Shoulder Press**

pages 126–127

❷ **Prone Arm Raise**

page 62

❸ **Cobra Stretch**

page 68–69

❹ **Swiss Ball Arm Flexion**

page 70

❺ **Swiss Ball Arm Extension**

page 71

❻ **One-Armed Row**

pages 36–37

❼ **Lateral Raise**

pages 40–41

❽ **Standing Fly**

pages 42–43

❾ **Standing Abdominal Brace**

page 58

❿ **Swiss Ball Pelvic Tilt**

page 64–65

⓫ **Lying Pelvic Tilt**

page 59

⓬ **Bridge**

page 60–61

TONING WORKOUT

Another all-around workout, this one aims at firming and enhancing the look of your muscles.

❶ Swiss Ball Incline Chest Press

pages 140–141

❷ Overhead Press

pages 38–39

❸ Dumbbell Upright Row

pages 142–143

❹ Biceps Curl

pages 44–45

❺ One-Armed Triceps Kickback

pages 46–47

❻ One-Legged Downward Press

pages 54–55

❼ Lunge

pages 134–135

❽ Swiss Ball Hamstrings Curl

pages 146–147

❾ Swiss Ball Jackknife

pages 96–97

❿ Swiss Ball Plank with Leg Lift

pages 88–89

⓫ T-Stabilization

pages 92–93

PERFORMANCE WORKOUT

Here's a workout designed to make you stronger and capable of powerful, accurate movements.

LOWER-BODY WORKOUT

With its focus on strengthening your abs, glutes, and legs, this workout will get you ready for the summer.

1 Squat

pages 48–49

2 Split Squat with Curl

pages 50–51

3 Swiss Ball Jackknife

pages 96–97

4 Leg Raise

pages 104–105

5 Reverse Crunch

pages 102–103

6 Swiss Ball Hip Crossover

pages 110–111

7 Sumo Squat

pages 132–133

8 Lunge

pages 134–135

9 Stiff-Legged Deadlift

pages 136–137

8 Swiss Ball Hamstrings Curl

pages 146–147

11 Dumbbell Calf Raise

pages 138–139

12 Side Lunge

pages 144–145

RANGE OF MOTION WORKOUT

Try this workout to noticeably improve your flexibility and reach.

❶ Chest Stretch

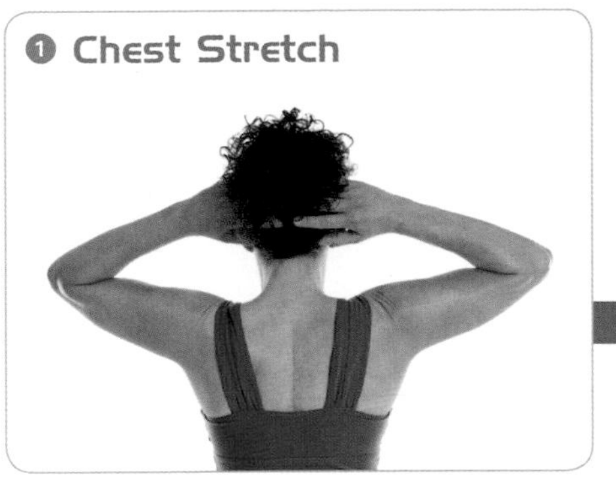

page 24

❷ Swiss Ball Kneeling Lat Stretch

page 16

❸ Standing Quadriceps Stretch

page 17

❹ Standing Hamstrings Stretch

page 20

❺ Shoulder Stretch

page 23

❻ Triceps Stretch

page 22

❼ Standing Biceps Stretch

page 25

❽ Back Pat & Rub

page 75

❾ Forward Arm Reach

page 76

❿ Sideways Arm Lift & Cross

page 77

⓫ Knee Lift

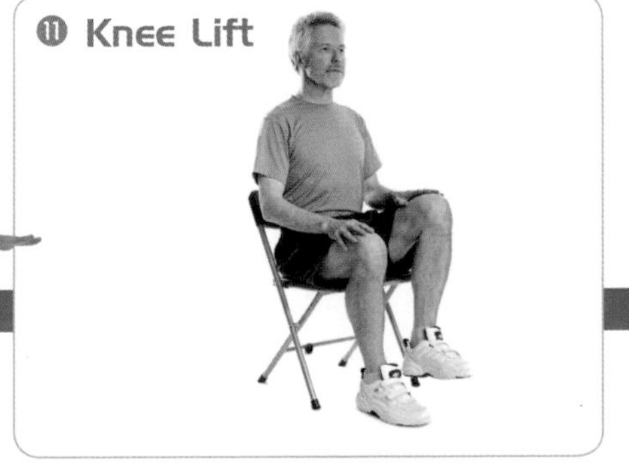

page 80

⓬ In-Out Rotation

page 81

GLOSSARY

GENERAL TERMS

abduction: Movement away from the body.

adduction: Movement toward the body.

anterior: Located in the front.

cardiovascular exercise: Any exercise that increases the heart rate, making oxygen and nutrient-rich blood available to working muscles.

cardiovascular system: The circulatory system that distributes blood throughout the body, which includes the heart, lungs, arteries, veins, and capillaries.

core: Refers to the deep muscle layers that lie close to the spine and provide structural support for the entire body. The core is divisible into two groups: major core and minor core muscles. The major muscles reside on the trunk and include the belly area and the mid and lower back. This area encompasses the pelvic floor muscles (levator ani, pubococcygeus, iliococcygeus, pubo-rectalis, and coccygeus), the abdominals (rectus abdominis, transversus abdominis, obliquus externus, and obliquus internus), the spinal extensors (multifidus spinae, erector spinae, splenius, longissimus thoracis, and semispinalis), and the diaphragm. The minor core muscles include the latissimus dorsi, gluteus maximus, and trapezius (upper, middle, and lower). These minor core muscles assist the major muscles when the body engages in activities or movements that require added stability.

crunch: A common abdominal exercise that calls for curling the shoulders toward the pelvis while lying supine with hands behind the head and knees bent.

curl: An exercise movement, usually targeting the biceps brachii, that calls for a weight to be moved through an arc, in a "curling" motion.

deadlift: An exercise movement that calls for lifting a weight, such as a barbell, off the ground from a stabilized bent-over position.

dumbbell: A basic piece of equipment that consists of a short bar on which plates are secured. A person can use a dumbbell in one or both hands during an exercise. Most gyms offer dumbbells with the weight plates welded on and poundage indicated on the plates, but many dumbbells intended for home use come with removable plates that allow you to adjust the weight.

extension: The act of straightening.

extensor muscle: A muscle serving to extend a body part away from the body.

flexion: The bending of a joint.

flexor muscle: A muscle that decreases the angle between two bones, as bending the arm at the elbow or raising the thigh toward the stomach.

fly: An exercise movement in which the hand and arm move through an arc while the elbow is kept at a constant angle. Flyes work the muscles of the upper body.

free weights: Any weight not part of a machine, including dumbbells, barbells, medicine balls, sandbells, and kettlebells.

hand weight: Any of a range of free weights that are often used in weight training and toning. Small hand weights are usually cast iron formed in the shape of a dumbbell, sometimes coated with rubber or neoprene for comfort.

iliotibial band (ITB): A thick band of fibrous tissue that runs down the outside of the leg, beginning at the hip and extending to the outer side of the tibia just below the knee joint. The band functions in concert with several of the thigh muscles to provide stability to the outside of the knee joint.

incline: Position in which the body is tilted back in respect to the vertical plane. Often used to work the upper-chest muscles.

lateral: Located on, or extending toward, the outside.

medial: Located on, or extending toward, the middle.

medicine ball: A small weighted ball used in weight training and toning.

neutral position (spine): A spinal position resembling an S shape, consisting of a lordosis in the lower back, when viewed in profile.

posterior: Located behind.

press: An exercise movement that calls for moving a weight or other resistance away from the body.

range of motion: The distance and direction a joint can move between the flexed position and the extended position.

reclining press: An exercise movement that calls for a person to lie supine on a bench or Swiss ball, lower a weight to chest level, and then push it back up until the arm is straight and the elbow locked. Reclining presses strengthen the pectorals, deltoids, and triceps.

resistance band: Any rubber tubing or flat band device that provides a resistive force used for strength training. Also called a "fitness band," "stretching band," and "stretch tube."

rotator muscle: One of a group of muscles that assist the rotation of a joint, such as the hip or the shoulder.

scapula: The protrusion of bone on the mid to upper back, also known as the "shoulder blade."

split squat: An assisted one-legged squat where the nonlifting leg is rested on the floor a few steps behind the lifting leg, as if it were a static lunge.

squat: An exercise movement that calls for moving the hips back and bending the knees and hips to lower the torso and an accompanying weight, and then returning to the upright position. A squat primarily targets the muscles of the thighs, hips and buttocks, and hamstrings.

Swiss ball: A flexile, inflatable PVC ball measuring approximately 14 to 34 inches in circumference that is used for weight training, physical therapy, balance training, and many other exercise regimens. It is also called a "balance ball," "fitness ball," "stability ball," "exercise ball," "gym ball," "physioball," "body ball," and many other names.

warm-up: Any form of light exercise of short duration that prepares the body for more intense exercises.

weight: Refers to the plates or weight stacks, or the actual poundage listed on the bar or dumbbell.

LATIN TERMS

The following glossary explains the Latin terminology used to describe the body's musculature. Certain words are derived from Greek, which is indicated in each instance.

CHEST

coracobrachialis: Greek *korakoeidés*, "ravenlike," and *brachium*, "arm"

pectoralis (major and minor): *pectus*, "breast"

ABDOMEN

obliquus externus: *obliquus*, "slanting," and *externus*, "outward"

obliquus internus: *obliquus*, "slanting," and *internus*, "within"

rectus abdominis: *rego*, "straight, upright," and *abdomen*, "belly"

serratus anterior: *serra*, "saw," and *ante*, "before"

transversus abdominis: *transversus*, "athwart," and *abdomen*, "belly"

NECK

scalenus: Greek *skalénós*, "unequal"

semispinalis: *semi*, "half," and *spinae*, "spine"

splenius: Greek *spléníon*, "plaster, patch"

sternocleidomastoideus: Greek *stérnon*, "chest," Greek *kleís*, "key," and Greek *mastoeidés*, "breastlike"

BACK

erector spinae: *erectus*, "straight," and *spina*, "thorn"

latissimus dorsi: *latus*, "wide," and *dorsum*, "back"

multifidus spinae: *multifid*, "to cut into divisions," and *spinae*, "spine"

quadratus lumborum: *quadratus*, "square, rectangular," and *lumbus*, "loin"

rhomboideus: Greek *rhembesthai*, "to spin"

trapezius: Greek *trapezion*, "small table"

SHOULDERS

deltoideus (anterior, medial, and posterior): Greek *deltoeidés*, "delta-shaped"

infraspinatus: *infra*, "under," and *spina*, "thorn"

levator scapulae: *levare*, "to raise," and *scapulae*, "shoulder [blades]"

subscapularis: *sub*, "below," and *scapulae*, "shoulder [blades]"

supraspinatus: *supra,* "above," and *spina*, "thorn"

teres (major and minor): *teres*, "rounded"

UPPER ARM

biceps brachii: *biceps*, "two-headed," and *brachium*, "arm"

brachialis: *brachium*, "arm"

triceps brachii: *triceps*, "three-headed," and *brachium*, "arm"

LOWER ARM

anconeus: Greek *anconad*, "elbow"

brachioradialis: *brachium*, "arm," and *radius*, "spoke"

extensor carpi radialis: *extendere*, "to extend," Greek *karpós*, "wrist," and *radius*, "spoke"

extensor digitorum: *extendere*, "to extend," and *digitus*, "finger, toe"

flexor carpi pollicis longus: *flectere*, "to bend," Greek *karpós*, "wrist," *pollicis*, "thumb," and *longus*, "long"

flexor carpi radialis: *flectere*, "to bend," Greek *karpós*, "wrist," and *radius*, "spoke"

flexor carpi ulnaris: *flectere*, "to bend," Greek *karpós*, "wrist," and *ulnaris*, "forearm"

flexor digitorum: *flectere*, "to bend," and *digitus*, "finger, toe"

palmaris longus: *palmaris*, "palm," and *longus*, "long"

pronator teres: *pronate*, "to rotate," and *teres*, "rounded.

HIPS

gemellus (inferior and superior): *geminus*, "twin"

gluteus maximus: Greek *gloutós*, "rump," and *maximus*, "largest"

gluteus medius: Greek *gloutós*, "rump," and *medialis*, "middle"

gluteus minimus: Greek *gloutós*, "rump," and *minimus*, "smallest"

iliopsoas: *ilium*, "groin," and Greek *psoa*, "groin muscle"

iliacus: *ilium*, "groin"

obturator externus: *obturare*, "to block," and *externus*, "outward"

obturator internus: *obturare*, "to block," and *internus*, "within"

pectineus: *pectin*, "comb"

piriformis: *pirum*, "pear," and *forma*, "shape"

quadratus femoris: *quadratus*, "square, rectangular," and *femur*, "thigh"

UPPER LEG

adductor longus: *adducere*, "to contract," and *longus*, "long"

adductor magnus: *adducere*, "to contract," and *magnus*, "major"

biceps femoris: *biceps*, "two-headed," and *femur*, "thigh"

gracilis: *gracilis*, "slim, slender"

rectus femoris: *rego*, "straight, upright," and *femur*, "thigh"

sartorius: *sarcio*, "to patch" or "to repair"

semimembranosus: *semi*, "half," and *membrum*, "limb"

semitendinosus: *semi*, "half," and *tendo*, "tendon"

tensor fasciae latae: *tenere*, "to stretch," *fasciae*, "band," and *latae*, "laid down"

vastus intermedius: *vastus*, "immense, huge," and *intermedius*, "between"

vastus lateralis: *vastus*, "immense, huge," and lateralis, "side"

vastus medialis: *vastus*, "immense, huge," and *medialis*, "middle"

LOWER LEG

adductor digiti minimi: *adducere*, "to contract," *digitus*, "finger, toe," and *minimum* "smallest"

adductor hallucis: *adducere*, "to contract," and *hallex*, "big toe"

extensor digitorum: *extendere*, "to extend," and *digitus*, "finger, toe"

extensor hallucis: *extendere*, "to extend," and *hallex*, "big toe"

flexor digitorum: *flectere*, "to bend," and *digitus*, "finger, toe"

flexor hallucis: *flectere*, "to bend," and *hallex*, "big toe"

gastrocnemius: Greek *gastroknémía*, "calf [of the leg]"

peroneus: *peronei*, "of the fibula"

plantaris: *planta*, "the sole"

soleus: *solea*, "sandal"

tibialis anterior: *tibia*, "reed pipe," and *ante*, "before"

tibialis posterior: *tibia*, "reed pipe," and *posterus*, "coming after"

trochlea tali: *trochleae*, "a pulley-shaped structure," and *talus*, "lower portion of ankle joint"

CREDITS & ACKNOWLEDGMENTS

All photographs by Jonathan Conklin/Jonathan Conklin Photography, Inc.

Models: Elaine Altholz and Peter Vaillancourt

All large anatomical illustrations by Hector Aiza/3D Labz Animation India, with small insets by Linda Bucklin/Shutterstock

Acknowledgments

I would like to thank the heroic and selfless over-50s who gave to others often at cost to themselves, including my beloved grandmother Claire, Grandpa Joe, grandmother Audrey, and, especially, my parents, Dale and Bruce Liebman, who continue to give and keep the Core 4 alive and strong.

The author and publisher also offer thanks to those closely involved in the creation of this book: Moseley Road president Sean Moore; art director Brian MacMullen; editorial director Lisa Purcell; editor Erica Gordon-Mallin; designers Danielle Scaramuzzo and Terasa Bernard; and photographer Jonathan Conklin.